W9-DEK-169

Social Work Science

Social Work Science

IAN SHAW

Columbia University Press *New York*

Columbia University Press
Publishers Since 1893
New York Chichester, West Sussex
cup.columbia.edu

Library of Congress Cataloging-in-Publication Data
Names: Shaw, Ian, 1945– author.
Title: Social work science / Ian Shaw.
Description: New York : Columbia University Press, [2016] |
Includes bibliographical references and index.
Identifiers: LCCN 2015039270 | ISBN 9780231166409 (cloth : alk. paper) |
ISBN 9780231541602 (e-book)
Subjects: LCSH: Social service—Practice. | Social service. | Social work education.
Classification: LCC HV10.5 .S455 2016 | DDC 361.3—dc23
LC record available at http://lccn.loc.gov/2015039270

Columbia University Press books are printed on permanent and
durable acid-free paper.
This book is printed on paper with recycled content.
Printed in the United States of America

c 10 9 8 7 6 5 4 3 2 1

Cover design: Lisa Hamm

To my doctoral students,
present and past

Contents

Social Work Science

Introduction

No social work is free from science. The ambiguous implications of these words permeate this book.

Is social work a science? I believe for various reasons that will surface throughout the book that this is an unhelpful question. It may be helpful to some degree to say that social work is a science-based occupation. Brante concludes: “A profession obtains its status from a central base, that it is a truth regime. Because of its scientific base, a profession is the ultimate link to ‘truth’; there is no higher authority. This and only this is what makes professions unique” (2011, 19). But this is some way from saying that social work as such is a science. This book is called *Social Work Science* to capture the multifarious ways in which, for good or ill, social work is never free of science and to bypass the miserable connectors “and,” “in,” “of,” and even “for.”

I would not be thought a pessimistic naysayer. In general, I bring a view about science that in some but not all ways is like that of Weber. Science, he insisted in the gendered language almost universal at the time he was speaking (1919), demands a “strange intoxication.” “Without this passion . . . you have no calling for science and you should do something else. For nothing is worthy of a man unless he can pursue it with passionate devotion” (Weber 1948, 135).

But while enthusiasm is a prerequisite, no amount of enthusiasm, remarked Weber, "can compel a problem to yield scientific results." "Calculation" is also indispensable. Perhaps with tongue in cheek, recognizing how such aspects of science have been thought at different times to be lower-level scientific tasks, he advises that "no sociologist . . . should think himself too good, even in his old age, to make tens of thousands of quite trivial computations." But further, some "idea" has to occur if one is to accomplish anything worthwhile. "Normally such an 'idea' is prepared only in the soil of very hard work," but this is not always the case. "Ideas occur to us when they please, not when it pleases us," for example, "when taking a walk on a slowly ascending street." "Yet ideas would certainly not come to mind had we not brooded at our desks and searched for answers with passionate devotion" (Weber 1948, 136). "Passionate minds" are not incompatible with writing that is voiced with "more rumination than indignation" (Riesman and Becker 1984, vii).[1]

I should say something about my general approach, which is shaped by seeing science work as a form of social action marked by a complex interplay of choice, action, and constraint. In principle I am not opposed to the idea that science in its relation to social work may be expressed rationally and, I would possibly say, systematically. But this is not in any exhaustive sense the organizing purpose of the book, which frequently is more occasional, historical, and contextual. Augustine spoke of the doctrine of the Trinity as a fence around a mystery—that is, it tells us what it is not and not quite what it is. I might say likewise of this book. I am trying at best to offer not an entire body but a skeleton without which the social work science body will be "flabby and misshapen." I am not offering incorrigible assertions. For example, I distinguish among evidence, understanding, and justice, but I do not seek to separate them. I am engaged much in defining, distinguishing, and connecting. In so doing I do not aim for some residual shared ground, a lowest common denominator. This I find of little use. It would be a poor social work science that contented itself with distilling what all statements of social work science have in common. While it is of undoubted use to identify what social work science in its varied manifestations has in common, it is equally important to identify and describe its particularities. Thus, I do not expect to find in a single or even two or three statements some way of capturing the essential meaning of social work science.[2]

This is not a methods book—or even a methodology book. There is almost no discussion of methods, except when I consider the history of methods or when I turn to questions of methodology for the illumination they shed on social work science. I take in one sense a conventional position in that I believe good practice—scientific or otherwise—is founded on good understanding. But where I place my emphasis in this book is on the argument that prescriptions for practice—whether evidence based, hermeneutic, postmodern, or whatever—too often leap to practice prescription with undue attention to understanding. So I spend perhaps more space on understanding the various forms of scientific practice than on starting from a "pure" statement of positions or being *for* or *against* any specific position. Of course I *am* critical of some central tenets and application, as I see them, of scientific social work in the United States and elsewhere. Am I "for" science? Taking the golem—the creature of Jewish mythology—as a metaphor for science, Harry Collins and Trevor Pinch, in their captivating book, seek "to explain the golem that is science. We aim to show that it is not an evil creature but it is a little daft. Golem Science is not to be blamed for its mistakes; they are our mistakes. A golem cannot be blamed if it is doing its best. But we must not expect too much. A golem, powerful though it is, is the creature of our art and our craft" (1998, 2).

In a sadly overlooked book from a half-century ago, Peter Nokes considered the ambiguity of attitudes to information in the welfare professions. "Customary ideas assure us that the search for truth is everybody's concern, yet in fact attitudes to truth, to *fact*, are highly ambiguous" (1967, 47). Speaking of England, but in ways that echo round the chambers of international social work, he says that "the possession of the right sentiments has always, except for a short period where the Utilitarian philosophy held sway, attracted more approval than the ability to pursue a train of thought" (48).

Why is the practitioner—and the social scientist, he will go on to say—unable "to maintain an entirely candid attitude to facts" (49)? For two reasons, he suggests. On the part of the practitioner, it is "the unacknowledged need to be reassured that all is well," while on the part of the social scientist, it is "the pressure to be committed to programmes of reform, to change things" (49). We might concur that to say that "convictions are one thing and science quite another has really no cogency in the social sciences" (65). We will have cause to return to the relationship between moral conviction and science several times in this book, especially toward the closing chapters.

The book may come to be seen as distinctive in several ways. *Historical* questions play a larger part in it and a different role than in most social work writing. In addition to chapter 3, I draw throughout the book on archival sources. I treat many problems in part through a historical lens. Consistent with my unhappiness with the persistence of Comtean views of scientific progress, I am as likely to find persuasive a writer from fifty or a hundred years ago as one from the present decade. Second, although the book is about *social work* science, it will soon be apparent that this in no way privileges social work writing. There is as much of Becker, Campbell, Collins, Cronbach, Giddens, Kuhn, Schutz, Weber, and Williams as there is of social work writers. This is not simply to widen and enrich social work—although it does *both* of these—but because many of the questions considered in the book do not originate in social work. While I do hope to convey a "social work imagination" and "gaze," for me the relationship between social work and, for example, sociology is a permeable membrane and not a wall.[3]

Third, the book deals in some detail with themes that either are absent from or treated in thin and rather unreflective ways in social work literature. For example, it is part of the platform of this book that one cannot understand social work *science* without delving in some depth into broader forms of knowledge and action. I have much to say about the meaning and implications of saying we have tacit knowledge, expertise, or commonsense understanding of something. I also give considerable attention to the nature of social work science as a social project, through elaborations of social work inventions, networks, controversies, and so on. The extent to which the voices of mainstream scientists are heard through the chapters is also perhaps original in social work writing.

Finally I have much to say about language. I should recognize here the unavoidable presence of gendered language. You will discover that I continue to "apologize" from time to time, but I was more surprised than perhaps I ought to have been to encounter the almost universal use of "men" and "he" across the whole spectrum of social work and social science writing, by both men and women, until relatively recently. I explore aspects of the substance of this issue in chapters 6 and 8.

Various ways will come to the fore in this book in which encountering both social work and science will seem like navigating between Scylla and Charybdis. The first two chapters, to continue the metaphor, can be read as mapping the various rock shoals and whirlpools. By way of introduction, we glance at the inheritance bequeathed us before asking whether it makes sense to think of social work as a science. Calling the opening chapter

"Talking Social Work Science" reflects the core themes and questions of this chapter. We will reflect on several key terms in science language. Following that, we ask in what general way social work science yields—or does not yield—a knowledge foundation. That takes us on to where we ask if social work science has a unity to it or if it is made up of several different and perhaps incommensurable enterprises that are fundamentally unalike.

Asking if there is one or more social work science leads us on to the opening part of chapter 2, in which we think of "Doing Social Work Science" as requiring us to think about how science is related to other kinds of knowledge and action. We will ponder in broad terms how science is related to personal knowledge, common sense, power, action, politics, and faith. In the subsequent parts of chapter 2 we will consider the practice of doing science before asking what is meant when we talk of doing *good* social work science.

These opening themes frame the following seven chapters. Science as part of social work—as something from which no social work is free—is in every part of social work's inheritances and will be central to its legacies. The early confidence in science both stimulates and troubles social work today. Bertrand Russell bemoaned "a too exclusive emphasis on the past" that engenders "a habit of criticism towards the present." He tellingly exemplified a tendency to disdain the past:

> The qualities in which the present excels are qualities to which the study of the past does not direct attention. In what is new and growing there is apt to be something crude, insolent, even a little vulgar, which is shocking to the man of sensitive taste; quivering from the rough contact, he retires to the trim gardens of a polished past, forgetting that they were reclaimed from the wilderness by men as rough and earth-soiled as those from whom he shrinks in his own day.
>
> (Russell 1913)

Social work's inheritance is explored in chapter 3. Chapter 4 looks at science and technology—a kind of knowledge and practice that sometimes is seen as very close to scientific practice. The issue of whether there are one or several social work sciences lies behind much of the book, but perhaps especially the first two chapters and, where we look at debates and disagreements, chapter 5. Chapters 6 through 8 elaborate three key motifs in social work science—science as seeking evidence, as gaining understanding, and as enhancing social justice. The final chapter and the appendix on "Writing Social Work Science" are, in different ways, both about the consequences of social work science. Thought of more generally, the book as a whole falls into two parts. In chapters 1 through 5 a series of general questions is pursued. What language and ideas form social work science? What is entailed in the doing of science in social work? How should it form our practice when seeing social work as a moment in past, present, and future time? How is science related to technology? What should we make of debates, disagreement, controversies, and inventions in social work? Chapters 6 through 9 unpack claims made earlier in the book.

I have done my best not to fall into the trap, in Foucault's words, of joining "the great warm and tender Freemasonry of useless erudition" (quoted in O'Farrell 2005, 87). The main way I have addressed this is through appending "Taking It Further" postscripts to every chapter. I suspect I have always ignored lists of "further reading" at the end of book chapters, so I offer suggestions for each chapter with some hesitation. I include references that are more of a "starter" reading and other readings that supplement the chapter at a similar level. "Taking it further" always entails an action, sometimes individual and other times jointly with others.

This has been a demanding book to write. Jonathan Franzen, when speaking of influences on his writing, observed: "As a writer . . . you owe it to your readers to set yourself the most difficult challenge that you have some hope of being equal to."[4] Foucault remarked, "The reader can easily tell when you have worked and when you merely talk off the top of your head" and added, "to work is to undertake to think something other than what you thought before" (O'Farrell 2005, 45). Having attempted to do so, and to write with critical good manners, I leave the reader to "tread on my dreams" because, to quote Foucault once more, "he who writes does not have the right to give orders as to the use of his writings" (O'Farrell 2005, 55).

[1]

Talking Social Work Science

This opening chapter focuses on the language of social work science. It sets out the range of problems and positions taken in relation to them and offers preliminary responses.

I open by asking what we should know and think about the language of science in social work. I take the idea of science being about understanding "nature" and then the meanings of "theory" and its relation to "practice" as examples; I will also consider the word "science" itself.

I ask if science can be seen in some senses as a single field, including how we should understand claims that science is or should be value free and "objective."

Following a generally historical account of how these themes emerged, we then trace the meaning of "positivism" and the emergence of positions that broadly may be called "postpositivist." This leads us into the meaning and significance of Kuhn's contribution to understanding science and thence to a brief outline of social constructionist positions in science.

This chapter and the one that follows should be seen as groundwork for all that follows in the book.

8

Talking Social Work Science

> Science is a grand thing when you can get it; in its real sense one of the grandest words in the world. But what do these men mean, nine times out of ten, when they use it nowadays? When they say detection is a science? When they say criminology is a science? They mean getting *outside* a man and studying him as if he were a giant insect; in what they would call a dry impartial light. . . . They mean getting a long way off from him. . . . When the scientist talks about a type, he never means himself, but always his neighbour; probably his poorer neighbour.
>
> —G. K. Chesterton, *The Secret Life of Father Brown*

"Science" frequently gets bad press in social work. Apart from those affronted by the very suggestion that Chesterton knew something about science, social workers are likely to warm to his famous character Father Brown's subsequent remark that "I don't try to get outside the man. I try to get inside. . . . I wait till I know I am inside a murderer, thinking his thoughts, wrestling with his passions." This sensitivity traces a path through social work over time. It is not at all unlike, for example, the nineteenth-century London housing reformer Octavia Hill, who apparently concluded, "By character is meant . . . knowledge of the passions, hopes and history of people; where the temptation will touch them, what is the little scheme they have made of their lives, or would make, if they had encouragement; that training long past phases of their lives may have afforded; how to move, touch and teach them." I say "apparently" because this is the Chicago sociologist Ernest Burgess echoing Mary Richmond's appreciation of Hill (Burgess 1928, 525).[1] Burgess's readiness to hear what social workers thought also can be seen when he asked Lewis E. Lawes, a Chicago social work figure who referred to his "twenty-four years of active prison work," to comment on a paper he had written on predicting parole outcomes. Lawes's comments included "When you deal with the problems of social service, in all its aspects, you are confronted with human emotions that will not lend themselves to cold scientific analysis" and a quotation from Burgess speaking of social workers who "sincerely feel that their services like those of religion are in the realm of intangible rather than material values and are not to be subject to crude measurement of statistical procedure" (Burgess Papers, Box 195). Four decades later, the British social work scholar Elizabeth Irvine bemoaned that "science deals splendidly with all that can be weighed, measured and counted, but this involves excluding from the universe of discourse the intangible, the imponderable, all that cannot be reduced to statistics" (Irvine 1969, 4).

What Should We Know and Think About the Language of Science in Social Work?

It ought, now and then, to bring us up short to realize that our language and the frames and terms we use draw on more general culture, that they go back to earlier times, or that they have at other points been seen very differently. In other words, our language about science has not only an intellectual but also a historical shape. "Academic subjects are not eternal categories" (Williams 1983, 14), and their language often exhibits "one culture chafing against another" (Hitchings 2008, 7). As Hitchings remarks in his enticing book *The Secret Life of Words,* "words contain the fossils of past dreams and traumas" and so invite "an archaeology of human experience" (2008, 4). Take, for example, "eugenics." "Darwin's polymathic cousin Francis Galton came up with *eugenic* in 1883; the politics of Social Darwinism were made respectable by means of a handsome Greek name" (274). This illustrates how "a more self-inspecting attitude tends to call for, or give rise to, a vocabulary more clearly touched by science—or by the illusion of scientific nicety" (275), such as in the "vehement obscurity" of parts of sociology (325). In these ways "language betrays frailties, anxieties and the precariousness of self-image" (325).

We use words to bind together certain ways of seeing society and social work. But we may also use words to open up issues and problems of which we need to be much more conscious. But so doing does not mean that the problems will thereby be solved. Understanding the complexities of the word "class," for example, does little to resolve class disputes and conflicts, just as understanding evidence-based practice or postmodernism does little to improve social work. But these disputes cannot be thought through—or even brought into focus—unless we are conscious of words as elements in the problems. Such an exercise can contribute, if not resolution, then, in Raymond Williams's words, "just that extra edge of consciousness" (1983, 24). There is much value in efforts to *understand* on their own terms positions about which one may have deep concerns and so to resist that tendency to "call all sects but our own sectarian" (16). Social work writing too often betrays a rush to evaluative assessments based on assumptions about the relative importance and meaning of scientific practice, postmodernism, or human understanding—"the rash and polemical extension from a proposition to a recommendation," as Williams nicely expresses it (199). What is needed is first to understand.

The vocabulary I am interested in is not only specialist social work science language but also strong, difficult, and persuasive words in everyday usage and words that began in a specialized context but have become "naturalized" within social work.[2] We should also be sensitive to words that travel in the opposite direction from everyday language and have been taken up and given specialized meanings. The field of technology has many of these.

There may still be echoes in social work of the response that says it is basically a matter of education and similarly that when we see a word the first thing we need to do is to define it—a product, Williams suggests, of the tradition and influence of defining dictionaries and classical education (for example, "What is the Latin root?"). For many words, that does help, but for words that involve ideas and values, we need to start from how such words have a history and complexity of meaning—for example, "nature," "rational," "subjective." This venture is not without problems. If we want to understand the vocabulary of, for example, "sociology," the information is fairly complete, by which I mean it can largely be found in written form. To *some* degree this is true of "social work" and perhaps of words like "psychology," but even there—and much more so for words like "scientific" or "practice"—they cannot be traced without recognizing the central location of the *spoken* language. A glance at the limited but interesting oral history of social work will quickly show this to be so.

It may now be less common to speak of the "proper" or "strict" meaning of a word, in the face of the modish views that a word means only what it is now taken to mean or that meanings are defined by contexts of use. Neither stance is very helpful. On the latter position, context, of course, is hugely important—I have argued this fairly radically for social work research (Shaw 2010). But "the problem of meaning can never be wholly dissolved into context" (Williams 1983, 22). Words get adapted, extended, transferred, altered, and even reversed[3]—and always these are not final.

Still by way of preamble, in this context we owe ideas of discourse to Foucault. The extracts from and summary notes in example 1.1 illustrate Foucault's careful, even tentative approach.

EXAMPLE 1.1

Foucault on Discourse

Among the overused and probably under-understood words that have wide currency in social work science, "discourse" has high standing. Foucault lamented the "equivocal meaning of the term *discourse*, which I have used and abused in many different senses" (120). He was always aware that it was a work in progress.

He defines discourse as "a group of statements in so far as they belong to the same discursive formation." However, discourse "does not form a rhetorical or formal unity, endlessly repeatable, whose appearance or use in history might be indicated." It is made up of "a limited number of statements for which a group of conditions of existence can be defined." By this he refers to fields of discourse and speaks of clinical discourse, psychiatric discourse, the discourse of natural history, and so on. He wishes to use the term in this way "to reveal a descriptive possibility" (121).

So discourses are made up of statements and are practices that follow certain rules. "A statement belongs to a discursive formation as a sentence belongs to a text, and proposition to a deductive whole" (130).

Discourse in this sense is not an ideal, timeless form. The problem is not to ask oneself how and why it was able to emerge and become embodied at this point in time; it is, from beginning to end, historical—"a fragment of history, a unity and discontinuity in history itself, posing the problem of its own limits, its divisions, its transformations, the specific modes of its temporality rather than its sudden irruption in the midst of the complexities of time" (131). It is not a document but a "monument," and it is not something to be interpreted. In this connection, he remarks, "it refuses to be 'allegorical' " (155). Rather like Kuhn's apparent understanding of a paradigm, one cannot easily stand outside it. Its "archaeology" will demand much.

Hence we should not think of discursive practices as *psychological* (for example, a conscious formulation of an idea or desire), or as *rational* (an inference), or as a case of more or less *competent speaking* (grammatical), but as statements falling within a discursive formation.

Source: Foucault (2002).

There are examples of Foucauldian analysis in social work, perhaps especially in work on the history of social work and science (e.g., Shaw 2015a, Skehill 2007).

While there are sometimes severe theoretical difficulties,[4] what are the "big" words we think of when we talk about science? They include "knowledge," "theory," "objectivity" (and "object," "objectivize," and so on), "subject," "subjective," "subjectivism," "nature," "naturalism," "reason," "rational" (and "rationalism," "rationalize," "rationalization," and "reasonable"), "experience," and "science" itself.

In the seventeenth century "science" began to be distinguished from "art," though not at all in the way we make that distinction now. Science then was regarded as a kind of *argument* rather than a kind of *subject*, thus foreshadowing some sense of how we use the word "research" and also how the word "analysis" was used in early U.S. social work. Our use of science as particular fields of study had still to appear. "Analysis" of research data was rarely spoken about in early American sociology and social work, and even when the word appears usually it carried the more general meaning of taking an intellectually strategic approach to research and writing. This is evident in a minute on publication policies regarding the early Chicago sociology monographs, which reads: "The grade of excellence of the published work will depend quite as much upon the care with which the studies are subjected to critical analysis and the materials revised and reorganized, as upon original planning of the investigation" (SSRC, Box 18, Folder 9). Breckinridge referred in a similar tone to how "the successful use of . . . records should produce a habit of thought . . . a power of analysis of community relationships" (1924/1932, 4).

A key differentiation came from elsewhere in the eighteenth-century distinction between experience and experiment. Changes in the idea of "nature" encouraged the development of the idea of experiment toward the external world, and the conditions for the emergence of science as the theoretical and methodical study of nature were then complete. The idea of the neutral methodical observer and external object of study became generalized to associate not only to science but also to "fact," "truth," and "reason" or "rationality," and this had profound consequences for other areas of study. The specialization of the word "science" "is perhaps more complete in English than in most comparable languages" (Williams 1983, 278), which causes difficulties especially between English and French. There are also continuing difficulties in our use of "scientific" to mean methodological rigor or the old and continuing sense of a demonstrative proof in an argument.

Scattered reflections on most of the list of "big" words occur throughout the book, for example, "knowledge" almost everywhere, "objective" here and in chapter 6, and "subjective" in chapter 7. Other words surface from time to time—"paradigm" and "positivism" later in this chapter. Here I say something about "theory" (a much-used social work term), "practice," and, albeit briefly, "nature."

Nature

Many social scientists, including some in social work, "appear to be believers in Natural Law insofar as they appear to believe that what is ethically desirable can only be desirable to the extent that it is rooted in empirical investigation; more familiarly that investigation can itself give us guidance about what we ought to do" (Nokes 1967, 79). There are Christian roots in this position in Francis Bacon and others (Hooykaas 1972), albeit ones that, even when recognized today, may not be accepted. It is a shoot from a root belief that reason and ethics alike stem from the same divine order and that therefore there cannot be any inherent conflict between them. But contemporary outworkings of this position are potentially unhelpful. The connection may appear in social work when, for instance, the assumption seems to be held that because a certain approach to intervention is a good thing, then it would be found to be effective only when properly tested. It is probable that the heated critical responses to 1960s intervention research in the United States and the United Kingdom, which seemed to show that social work was not demonstrably effective, were driven in part by this confusion of desirability and effectiveness. The suggestion that adherence to demonstrably effective practice should be written into sanctionable codes of ethics encounters a difficulty that to this writer appears intractable—"that only a moral argument can be advanced for some of the policies that enlightened opinion would approve" (Nokes 1967, 81). For not every policy or practice aimed at the well-being, support, or change of those with or for whom practitioners work is measurably effective. The only justification left for such practices is "that they represent ways of dealing with people that we feel to be right and proper" (81). Anyone who thinks this dilemma is merely a transitional feature is living in a different universe.[5] We will see in chapter 4 that it is difficult to believe that technological applications of science could transform social work into a purely instrumental form of action.

Nature is often set against "society," so that efforts to avoid polarizing positions are often evident. Latour, for example, remarks that what at first was a distinction (Nature/Society) "becomes a separation, then a contradiction, then an irreconcilable tension, then an incommensurability, to end up in compete estrangement" (1992, 20). He wants to explain both Nature and Society in the same terms, by speaking of facts as "what is at once fabricated and what is not fabricated by anyone" (9). Facts are both object and subject, Nature *and* Society. He speaks of two realisms—social and natural—and of their being one and the same. So he would not want us to speak of, for example, evidence-based practice as half-natural and half-social. He would say it is neither object nor subject but a "quasi object." "It is out of their production and circulation that something originates that looks like Nature 'out there,' as well as something like Society 'up there' " (11).

Theory and Practice

"Theory" carries a range of meanings and a significant distinction from—and later opposition to—practice. From Aristotle we receive the distinction between theoretical reason, which concerns what we ought to believe, and practical reason, which concerns what we ought to do. Without going further back, by the seventeenth century "theory" had an array of meanings including that of a spectacle (a sight), a contemplated sight, a scheme of ideas, and an explanatory scheme (Williams 1983). So "theory" was sometimes used interchangeably with "speculation," but sometimes in contrast. Theory is always used in relation to practice as an interaction between things done, things observed, and the systematic explanation of these. Yet while this requires a distinction, it does not require opposition. We observe this when Popper insists, "Practice is not the enemy of theoretical knowledge but the most valuable incentive to it" (1966, 222). We also notice it when Geertz remarks that theorizing is difficult. It requires "the need for theory to stay rather closer to the ground than in other sciences. Only short flights of thought work. The tension between the pull of . . . the need to grasp and the need to analyse is . . . both necessarily great and essentially irremovable" (1973, 24). "Practice" as conventionally used also takes us to the whole area of kinds of knowledge—for example, tacit, common sense, and routine—matters to which we return in chapter 7.

The word *praxis* is used to suggest a new relation of theory and practice. The sense we give it is from the Marxist/Hegelian notion of "practice informed by theory," intended to unite theory in the sense of explanation and a scheme of ideas with a strong sense of the "practical," that is, practice in action. "Praxis" carries the idea that theoretical and practical elements can be distinguished but that it is always a whole activity and is to be judged as such. "The distinction between theory and practice can then be surpassed" (Williams 1983, 318).

Holding discussion of positivism for a moment, Phillips (1990a) provides a plausible account of the reasons for the demise of positivism in ways that help draw out important questions regarding theory in social work science. Three developments have been crucial in creating near unanimity among social scientists that there are no absolute justifications of scientific assertions. First, the role of observation as the final arbiter has been reappraised. For example, the acceptance that some mechanisms are unobservable led to the rejection of the belief that concepts can be reduced to a set of operational, observational statements. Equally influential has been the rejection of the assumption that observation can be theoretically neutral. The philosopher of science Hanson was important in developing this position and explicating how "there is more to seeing than meets the eyeball" (1958). Popper was equally vigorous.

> The positivist dislikes the idea that there should be meaningful problems outside the field of "positive" empirical science—problems to be dealt with by a genuine philosophical theory. He dislikes the idea that there should be a genuine theory of knowledge, an epistemology or a methodology. He wishes to see in the alleged philosophical problems mere "pseudo-problems" or "puzzles."
>
> (Popper 2002, 29)

Popper goes on to mock such a position. "For nothing is easier than to unmask a problem as 'meaningless' or 'pseudo.' All you have to do is fix upon a conveniently narrow meaning for 'meaning,' and you will soon be bound to say of any inconvenient question that you are unable to detect any meaning in it" (29). However, realist postpositivists of Popper's or Phillips's stances are united in arguing that theory-laden observation does not entail relativism and that it is possible to sustain a version of a correspondence view of truth—that research can represent social phenomena that are independent of it (Hammersley 1995, chap. 4; Phillips 1990b). Reid insisted that this was "not to argue for a dogmatic, elitist position. The argument is rather to make the most rational decisions possible under conditions in which uncertainty and error are more the rule than the exception" (1994b, 468).

Second, the relationship between theory and observation was shown to be more complex than previously thought. It became clear that theories are underdetermined by nature, such that we are never able to say that we have the best theory, and a variety of theories can be constructed that are equally compatible with the available evidence. This has led to the unanimous rejection of naïve foundationalism—the view that research findings of indisputable validity can manifest a direct correspondence with the real world. We should note, however, that this should not lead to a belief that theories are always fallible whereas the empirical base for them is not fallible—a trap that can be observed from time to time in social work writing. Lakatos (1970) labels this position "dogmatic falsificationism." He rightly rejects the caricature that "science grows by repeated overthrow of theories with the help of hard facts" (97). This skepticism regarding the value of theory simply reinstates a new version of theory-free observation. Observations are so in the light of a given theory, and "there are and can be no sensations unimpregnated by expectations" (99). However, although there are no *absolute* justifications, this does not mean there are *no* justifications. Yet while we may be warranted in holding particular views, we cannot assert that something is true, that our warrant is unchallengeable, or that it will forever be warranted—"*nothing* can guarantee that we have reached the truth" (Phillips 1990a, 43). There are ironic, invisible inverted commas whenever Popper speaks of "experiment," "empirical base," "observation," "applying theories," and so on. "If a theory is falsified, it is proven false; if it is 'falsified' it may still be true" (Lakatos 1970, 108).

Third, the view that it is the nature of science to develop through steady accumulation of findings and theories has been effectively challenged by Kuhn, Popper, Lakatos, and others.[6]

"Theory" is, of course, difficult to define. Indeed, we have already argued from the theory-laden character of observation that there is no hermetically sealed boundary between theory and fact. In addition, difficulties exist because theory may refer to formal propositions, a worldview, or a working hypothesis. "You can't pick up rocks in a field without a theory," as William

James apparently said. We often also think of "theory" as something big. Foucault may have thought in this way when he said: "Rather than *founding* a theory—and perhaps before being able to do so (I do not deny that I regret not yet having succeeded in doing so)—my present concern is to *establish* a possibility" (2002, 128–129). Yet those who claim to do atheoretical social work science are actually doing one or some combination of three other things:

> 1. They hold their theories tacitly, without reflecting upon them or stating them explicitly.
> 2. They hold them explicitly but deliberately withhold them from public view.
> 3. They pack structural concepts that properly belong to theory into their methodology where they are hidden from their view as well as ours.
> (Garrison 1988, 24)

How do ideas about "critical" theory fit? Critical theory is a catchall term to include neo-Marxist science, some feminist positions, the work of Paolo Freire, and some forms of participatory inquiry. These approaches are "critical" in the sense that problems are conceptualized as part of the social, political, and cultural patterns in which the subject of inquiry is formed. Paul Michael Garrett is among those who work on the implications of critical theory for social work practice (e.g., Garrett 2013). Critical inquiry focuses on the contradictions of practice. Hence the basic logic is preoccupied with "particular forms of reasoning that give focus to scepticism towards social institutions" (Popkewitz 1990, 49). This typically involves an "inversion" that entails "making history fragile . . . thus poking holes in the causality that confronts us in daily life and that limits our possibilities" (49).

Horkheimer, speaking in the 1930s, distinguished in this connection traditional and critical theory. For critical theory, the current social situation influences scientific structures and not "sheer logic alone" (Horkheimer 2002, 195). Ideas of the purpose and goals of research play their part, but these are not self-explanatory nor a matter of insight. Theorizing is not only "an intrascienfic process but a social one as well" (196). "In traditional theoretical thinking, the genesis of particular objective facts, the practical application of the conceptual systems by which it grasps the facts, and the role of such systems in action, are all taken to be external to the theoretical thinking itself" (208). This finds expression in the separation of fact and value, knowledge and action.

Critical thinking, on the contrary, "is motivated today by the effort really to transcend and to abolish the opposition between the individual's purposefulness, spontaneity, and rationality, and those work-process relationships on which society is built. Critical thought has a concept of man as in conflict with himself until this opposition is removed" (210). "In genuine critical thought explanation signifies not only a logical process but a concrete historical one as well. In the course of it both the social structure as a whole and the relation of the theoretician to society are altered, that is both the subject and the role of thought are changed" (211). "Traditional theory may take a number of things for granted: its positive role in a functioning society, an admittedly indirect and obscure relation to the satisfaction of general needs, and participation in the self-renewing life process. But all these exigencies . . . are called into question in critical thought" (216). Hence, critical knowledge is never neutral. "It is always for some particular subject" (Comstock 1982, 374). The production of knowledge is the production of values.

Critical theory does not exhaust the counterpoints to mainstream debates regarding theory. Understanding of the relationship of theory and practice has been shaped by a resurgence of interest in Aristotelian approaches to ethics, which are marked by deliberative reasoning and case-by-case decisions through a process of practical wisdom, *phronesis*. This has connections with pragmatism. This is not the same as a naïve "if it works it's true" position but is rather an antiessentialist approach.[7] Quoting Richard Rorty, Schwandt concludes, "When confronted with a knowledge claim the pragmatist is less concerned with whether it is right and asks instead, 'What would I be committing myself to?' . . . This shifts the focus of inquiry from verification and the appeal to method, to practice and an appeal to deliberation and conversation" (Schwandt 1993, 18).

Where does this take us? In the first place, we should not be precious about theorizing in general or our theories in particular. Too much time spent discussing theory is paralyzing. We need to defend social work against attacks and hence need those people who pursue "philosophical and methodological worry as a Profession" (Becker 1993, 226). But while we still have to do theoretical work, "we needn't think we are being especially virtuous when we do it" (221). There are some circumstances where theory has been overemphasized. For example, emancipatory researchers, especially some of those whose work is underpinned by neo-Marxism, should not feel too aggrieved by Lather's vigorous warnings against theoretical imperialism—the "circle where theory is reinforced by experience conditioned by theory" (Lather 1986b, 261). "Theory is too often used to protect us from the awesome complexity of the world," and "in the name of emancipation, researchers impose meanings on situations rather than constructing meaning through negotiation with research participants" (267, 265). Too often, "one is left with the impression that the research conducted provides empirical specificities for more general, *a priori* theories" (Lather 1986a, 76). Lather's remarks point to our second conclusion—that at their worst "scientific concepts can reinforce a vast array of dangerous or hateful political and moral agendas" (Jacob 1992, 495).

Third, we should not retreat behind the bunker of the threat of abuse by theory to protect ourselves against the demand that theory and critique must go hand in hand. The gap between theory and practice is typically formulated as a knowledge-transfer problem, in which practitioners fail to adopt research findings. Social workers often face the "scholastic fallacies" (Nowotny 2003, 156) sometimes evident in the dominance of social science and research "experts" over practice "beneficiaries." As Schön has expressed it, "Research and practice are presumed to be linked by an exchange in which researchers offer theories and techniques applicable to practice problems, and practitioners, in return, give researchers new problems to work on and practical tests of the utility of research results" (1992, 53).

The focus of Schön and other action scientists such as Argyris "have focused on the characteristics and behaviors of researchers to explain this lack of implementation of research knowledge. They argue that scientific knowledge will be implemented only if researchers, consultants, and practitioners jointly engage in interpreting and implementing study findings" (van de Ven and Johnson 2006, 804). Research and practice need linking in ways that release the potential for practice to challenge social work science and in so doing contest conventional hierarchical ways of seeing expert/beneficiary relationships. When theory/practice relationships are viewed as the failure of practitioners to apply the theories developed by those who are engaged in empirical and theoretical pursuits, this distorts reality. To regard theory and practice problems as breakdowns in communication that afflict practitioners is to fail to recognize that practical problems of this kind occur in the course of any theoretical undertaking. To assume that they can somehow be identified and tackled in theory and then "applied" in practice tends to conceal how they are generated out of practice. To translate and communicate are equally demanding. Film often illustrates this, in ways comedic (*Lost in Translation*) or dramatic (*Babel*). The tasks of translating and inhabiting exemplify that the relationship between the logics of doing social work and doing social work science is one of conjunction but difference.

Finally, our vision of the relationship between social work and science must never be utopian—but it must always be radical. Einstein may have said that "Things should be as simple as possible but not any simpler." Theory and research should persistently entice us with glimpses of the possibility of seeing the world differently. "The impact of theory will perhaps be greatest where the ideas have sufficient 'fit' with the issues professionals encounter on an everyday basis, while at the same time providing an alternative framework within which to understand these issues" (Brodie 2000).

This brings us full circle to the relationship of theory and practice, where "practice" has too frequently been the subservient partner. The problem stems to a significant degree from the low esteem afforded to manual labor and, in turn, of applied science. This goes as far back as Plato, who sought to separate experimental and applied science from pure science. "Only in medicine was manual work really honoured by the Greeks" (Hooykaas 1972, 80). Kirk and Reid (2002) hear echoes of this in some contemporary conceptions of evidence-based practice. Hooykaas approvingly quoted Thomas Spratt, the eighteenth-century author of *The History of the Royal Society of London*, saying, "philosophy will then attain to perfection, when either the Mechanic labourers shall have philosophical heads, or the Philosophers shall have mechanical hands" (Hooykaas 1972, 96).

Science and Sciences

Can we speak of the unity of science? Is there any sense in which foundationalist positions are plausible, in which we can be confident that the results of science have a strong correspondence to the real world? What do advocates and critics mean when they talk about positivism (and postpositivism)? Is any kind of positivism tenable following the work of Popper, Kuhn, and others? What might all this mean for how we think of social work outcomes and indeed causes more generally? What does social constructionism entail, and where does it take us?

In an article well regarded among advocates of scientific social work, Platt (1964) makes his case for the unity of the sciences and bases this on a logic of analytic induction that he calls "strong inference."

> Strong inference consists of applying the following steps to every problem in science, formally and explicitly and regularly:
>
> 1. Devising alternative hypotheses.
>
> 2. Devising a crucial experiment (or several of them), with alternative possible outcomes, each of which will, as nearly as possible, exclude one or more of the hypotheses.
>
> 3. Carrying out the experiment so as to get a clean result.
>
> Recycling the procedure, making sub-hypotheses or sequential hypotheses to refine the possibilities that remain; and so on.
>
> (347)

Without engaging in detailed textual exegesis, it should be noted for our purposes that Platt regards this procedure as applying to "every problem in science"—a hugely far-reaching claim and one, of course, that would dismiss as nonscience much that presently goes under the name of social work science. He also seems to assume that these steps are straightforwardly doable. An equally vigorous position espousing a single way of imagining science is given by Wallace. "Science is a way of generating and testing the truth of statements about events in the world of human experience" (2004, 35). Wallace contrasts science with modes he calls "authoritarian," "mystical," and "logico-rational"—a recurring theme we will encounter again in chapter 3. Compared with each of these, "the scientific mode combines a primary reliance on the observational effects of the statements in question, with a secondary reliance on the procedures (methods) used to generate them" (36). Reliance on observation—assumed to be at least partly independent of the observer—seeks "the annihilation of individual bias and the achievement of a 'universal' image of the way the world 'really' is" (37).

Donald Campbell remarked toward the end of his career that

> There may have been a past time in which our model of science confidently posited context-free, atomic, indubitable facts; timeless, context-free, unrestricted, covering laws and theories; and a possible

> language of unambiguous, monosemic, context-free words and grammars to describe these facts, laws and theories. If so, that time is well past.
>
> (1991, 587)

Yet it seems to suit critics of various hues to wish such a straw figure existed. Michael Billig, in pleading the merits of "the traditional, ill-defined skills of scholarship" against "rigorous, up-to-date methodology," characterizes the latter as involving rules of procedure that are "impersonal, in that they are meant to apply equally to all researchers. It is assumed that any two researchers who approach the same problem should arrive at identical results . . . Methodology attempts to standardize the practice of the social sciences and eliminate quirkiness" (2004, 13–14). Social work critics of alleged positivism sometimes adopt similar rhetorical devices.

Three related questions are raised by questions of the unity of science. How should we understand the relationship between the natural sciences and the social sciences? How should we understand claims that science is value free? What might we be saying if and when we speak of social work science as being "objective"?

There are, of course, several issues tangled up in the first question, some of which are open to empirical work:

1. Are the concepts and theories of social work of a similar or different order to those of the natural sciences?
2. Are the methods of social work research of a similar or different order to those of the natural sciences?
3. In what ways are the practices of scientific communities in social work, the social sciences, and the natural sciences comparable?
4. Should social work engage with the natural sciences and with natural scientists?

On the matter of value-free science, many of the consequent questions can be seen by scanning a long-ago discussion by Madge. At one extreme, he says, "is the stand taken by those who may be called naïve behaviourists, who are determined that all science is value-free" (1953, 14). He interestingly infers that "they not only deny the presence of their own intrinsic evaluations but are even very reluctant to admit that the feelings and beliefs of their human subjects are of concern to them, except when these are expressed in the form of overt acts" (14).

But the ideal goes far beyond behaviorists. Madge writes that Weber advocated the position "with particular subtlety" (14) in opposition to "professorial prophets" in Germany. He proposed a self-discipline aimed at ethical neutrality. "He regarded this as an unattainable ideal, but nevertheless one which should guide the social scientist" (15). Madge suggests that while there is no logical inconsistency on this position, it is "doubtful whether any human being is capable of the detachment that would be required of the true neutral" (15). "Freedom from value-judgment is an unreal imperative because it is unrealisable" (17).[8]

A further position is that the social scientist *ought* to have a standpoint on social issues and that without it the investigator would "lapse into the random collection of pointless facts and will claim immunity from the need to organise these facts *and to act upon them* until some unattainable time when all the data are safely gathered in" (15). This implies a duty to choose a position within certain limits and in accord with values of society. But which? These tend to be so abstract that "in practice they tend to disappoint, for they seldom offer useful guidance as to which particular norm or ethical principle should be emphasised in particular circumstances" (16), and conflicts of interest constantly occur. "It is just on such occasions that human ingenuity can justify or condemn—on rationalist and relativist grounds—practically any policy put forward by a sectional interest" (16). Madge comes down on the position not too far from one taken by Cronbach, which we will discuss subsequently, that the social scientist "may take on without shame the job of social engineer. . . . In this capacity he can goad his fellow citizens on to discontent . . . can help them choose by experiment between different ways forward" (17). This would, of course, be deeply at odds with the critical theorist.

Given Madge's eventual "value" position, it may not come as a surprise to hear him say "in no science is the pursuit of objective knowledge more futile than it is in social science. . . . To postulate an objective social science is to ask for something which is probably unattainable and may even be undesirable" (2, 6).

However, the term "objectivity" means different things to different people. Some would not be very far from a strict positivist position in respect of their apparent belief that research can provide an account closely approximate to reality. Others take the position that, while some version of objectivity remains as a "regulatory ideal" (Phillips 1990a, 43), social work science is always significantly jeopardized by interests, the social location of the inquirer and powerful stakeholders, and, in Guba's surprisingly realist phrase, "nature's propensity to confound" (1990, 19). Taking value-free knowledge and objectivity as normative ideals are for most purposes one and the same argument. The position is sometimes adopted by Christian philosophers of science (e.g., Helm 1987; MacKay 1981, 1987).[9] Thus MacKay readily concedes that "any idea that the *practice* of science can be value-neutral is nonsensical; our decisions whether, when and at what cost to lift the lid of Pandora's Box . . . are as value-laden as human judgements can be. But for the working scientist . . . the leap that he should on these grounds abandon his ideal of value-free knowledge as a 'myth' is a monstrous *non-sequitur*" (1987, 14). However, there remains an element of ambiguity as just to what this regulative norm commits.

Regardless of the faith position of the social work scientist, Helm's (1987) conclusions about why an objective *attitude* is a good thing are of wider value to most epistemological starting points. The nearest he comes to defining an objective attitude is "adopting the attitudes of impartiality, a willingness to have second thoughts, to consider alternative hypotheses, together with their associated practical techniques, such as careful measurement and the use of controlled experiments" (38).

1. It welcomes the checking of provisional findings. We might say "that although objectivity does not ensure investigative success it minimizes (while not eliminating) the effect of human fallibility and so minimizes the prospect of failure" (35).
2. It makes it "more possible to be *surprised* by the facts" (35).
3. It provides the means for settling disagreements.

He accepts that "objectivity presumes certain values" and that "a preference for objectivity as against subjectivity manifests still more values." However, "while there is nothing logically compelling about such values, it does not follow that there is *no reason whatever* to adopt some and shun others" (35). Behind both his first two points is an assumption of a more open mind. His argument also appears to assume that the opposite of being objective is to be liable to be the victim of prejudice and bias and liable to idiosyncrasies, hunches, and being swayed by first impressions—all terms Helm uses. In sum, he argues not that epistemic objectivity is sufficient for knowledge but

> because many human inquiries relate to a world of truth which are true irrespective of a person's own interest and feelings, objectivity in inquiry, though neither logically necessary nor logically sufficient for gaining beliefs which happen to be true, is usually necessary for the gaining of true belief, and is therefore desirable in any inquirer.[10]
>
> (39)

On subjective states he distinguishes statements such as "I feel miserable" and "I have depression." On most assumptions, the first statement is one that the utterer could not deny without being dishonest; the second is one that in principle is deniable. He is not saying that such statements have nothing to do with inner states. "To suppose this would lead us unacceptably in the direction of reductionist behaviourism and physicalism." Rather, a statement such as "I have depression" "is a complex utterance, with complex truth-conditions, to at least some of which evidence other than that which the speaker obtains directly about him or herself is relevant" (39).

While this take on things is likely to be congenial to a wide range of positions, those holding to a poststructural interpretivism of the form developed by Denzin would find it unacceptable. "In the social sciences there is only interpretation. Nothing speaks for itself" (Denzin 1994, 500). Hence "we do not study lived experience; rather we examine lived textuality," and "the direct link between experience and text . . . can no longer be presumed" (Denzin 1997, 33).

> Thus the mystery of experience. There is no secret key that will unlock its meanings. It is a labyrinth with no fixed origins and no firm center, structure, or set of recurring meanings. All that can be sought is a more fully grounded, multisensual, multiperspectival epistemology that does not privilege sight (vision) over the other senses, including sound, touch and taste.
>
> (1997, 36)

Schutz offers a final note on "the problem of the methodological unity of the empirical sciences" (1967e, 65). We should not look for this, he believes, "in a different logic governing each branch of knowledge" (65). But this does not mean conceding that only methods used by natural sciences are scientific. Rather than set up disciplines such as physics as providing a pure model of science, he poses "the question whether the methodological problem of the natural sciences in their present state is not merely a special case of the more general, still unexplored, problem how scientific knowledge is possible at all and what its logical and methodological presuppositions are" (66).

Positivisms, Postpositivism, and Thomas Kuhn

We have anticipated several related issues that now need unpacking. In roughly historical order, these relate to positivist science, the range of what are often called "postpositivist" standpoints, and the Kuhnian argument about science and scientific change.

Positivisms

In the nineteenth century positivism was in the ascendant, with its approach that knowledge, or at least all that should count as such, could be apprehended through the senses. The method of the natural sciences was seen as applicable to social phenomena. Its significance extended far beyond the field of science (cf. Burrow 1970, Shaw, 2014a). The decline of this faith took place along with a corresponding "reappraisal of traditional and habitual beliefs and modes of action" (Giddens 1993, 136). Logical positivism emerged as a reaction, as did phenomenology and linguistic philosophy as contrary positions. From 1900 to 1950 newer forms of positivism provided the dominant methodological justification of "normal science," "a science practised without constant reference back to fundamental philosophical premises" (Hughes 1980, 35). Positivism claims:

- Reality consists in what is available to the senses.
- An aversion to metaphysics. Religious, moral, or aesthetic statements were seen as without scientific meaning or at best as statements of preference and taste.

- Social and natural sciences share a common logical and methodological foundation, although this does not mean that they use the same methods of inquiry.
- The distinction between fact and value is fundamental, but this may not include a dualism of mind and matter.

Thus it is marked by a pervasive distinction between empirically grounded knowledge and mere speculation. The distinction between things human and things material is rejected at the level of science. In this way observational language was ontologically and epistemologically privileged. In positivism the interest lies in method rather than theory. For example, "cause" is treated empirically, and reason is seen as insufficient. The language of "variables" and other mathematical kinds of terms such as "operationalization," sampling, scaling, correlation, regression, and so on were central to this way of developing positivist visions (cf. Danziger and Dzinas 1997). These are ways of translating concepts into discrete chunks of data. "Objective data is produced by standardised instruments aimed at eliminating sources of bias of all kinds, and providing a neutral observational language" (Hughes 1980, 41–42).

We traced some of the reasons for the decline of positivism when discussing theory. Observation is very complex, and "facts" are not obvious. We are left with "methodological individualism" where "all social facts are in principle explainable in terms of facts about individuals" (Hughes 1980, 45). Durkheim expressed the classic positivist position that nothing about social facts is observable except their individual manifestations (Hughes 1980, 43).[11] This illustrates a fundamental problem of positivism in all versions when applied to social work: it has no plausible place for social wholes. But if there are no *observable* wholes, then positivism has no basis for arguing their existence. This is what happens when what is real is restricted to what can be observed.

This connects to a further problem in positivism: the inadequate status of social theory, regarding, for instance, causality. The classic response to this from positivist scholars is that "the method of interpretive understanding may be a useful adjunct to social science, as a source of 'hypotheses' about conduct, but that such hypotheses have to be confirmed by other, less impressionistic descriptions of human behaviour" (Giddens 1993, 61). As Thyer—among the many who could be quoted to this effect—expresses it, qualitative studies

> involving small numbers of individuals selected on the basis of convenience can generate hypotheses to be more rigorously tested on larger numbers of more representative persons, in order to develop generalizable conclusions and test theory. In contrast, quantitative research is not particularly well-suited to create hypotheses or to develop theories. Its strengths lay more in testing these, once formulated.
>
> (2012, 120)

In sum, the philosophical difficulties for positivism include "the problems of a neutral observation language, the relationship of theory to data, the matter of social wholes, and the issue of the scientific description of data" (Hughes 1980, 123).

Postpositivism

Popper's position was a radical break with logical positivism. Logical positivism sought to reduce meaning to testability. Popper substitutes the establishment of criteria of demarcation of science from other knowledge and the logic of falsifiability within a deductive framework.

The prefix "post-" is troublesome. Generally speaking it seems to convey a temporality—we used to have the one, but now we have the other. That idea is present in the phrase "late modernity," for example, which carries the idea, associated with writers such as Anthony Giddens and Zymunt Bauman, that modernity continues in ways that are manifest especially in society seen as "global." This argument is partly against ideas of postmodernism when they signal radical discontinuity with what it succeeds. In the case of "postpositivism" the relation with "positivism" is less clear. It shares a realist ontology but is fallibilist when it comes to epistemology. The difficulty arises from two problems. First, fallibilism covers a huge span of positions, ranging from cases where it may be relatively "weak," such that one remains confident that it is possible to gain a substantially true understanding of the world, to those whose fallibilism is relatively "strong," where confidence in the correspondence of findings of inquiry to the real world is generally rather low.

The second problem, which is both philosophical and logical, is how we simultaneously may hold ontological realism and epistemological fallibilism. The problem is not easily resolved, though I do not know of any social work argument where it is explicitly addressed. It holds that there is an objective reality that can be seen (known) by the perceiver, that such objects and relationships between them exist separately from the knower, and that knowledge of this reality is neither direct nor infallible but is "edited" by the objective referent and reflected in the convergence of observations from multiple independent sources of knowing. This was Campbell's position. He wedded epistemological naturalism to ontological realism.

- Epistemological naturalism: "the view that our knowledge *of* the world is itself a fact *in* the world and that the theory of knowledge may therefore be regarded as a part of empirical science" (Skagestad 1981, 77)
- Ontological realism: "the view that the world of which we have knowledge exists independently of our knowledge of it" (77–78).

He called this position one of "hypothetical realism" (or "critical realism")—"hypothetical" because it is an underjustified presumption. It fundamentally shaped his approach to arguments regarding paradigms and philosophical positions in science in general (Campbell 1991) and has similarities to subsequent work on critical realism. Some would defend the position through ideas about there being a drift toward scientific consensus that links the rules of replicability and consensual validation to ontological realism. However, it is possible to see these as shared biases, a common *theory* of reality rather than objective truth. The "problem is to disentangle shared perceptions from shared reality" (Brewer and Collins 1981, 4).[12]

Campbell sought to resolve the problem through a natural-selection model of knowing (Richards 1981). He drew on social evolutionary philosophy, whereby ideas are initially generated "blind" with no foresight as to their ultimate usefulness. He also applied this to cultural and moral traditions. Indeed, he said his own career exemplified the "blind-variation-and-selective-retention epistemology—now one of the core concepts in my theory of science" (Campbell 1981, 455). Helm (2014) starts from the same problem but seeks to resolve it through a Christian set of arguments and assumptions.

One of the most articulate defenders of postpositivism is Derek Phillips (1990a, 1990b, 2000), and we outlined his key departures from positivism in discussing the meaning and place of theory. Is this a weaker form of positivism? Much depends on one's starting point. For ontological realists it is a major departure, but for relativists it is a shift in degree rather than kind. For Phillips "in no sense is (this) new philosophy of science . . . closely akin to positivism" (1990a, 39). Is it based on an outmoded doctrine of realism? Here—and also in response to arguments that there are multiple and constructed realities—Phillips focuses on distinctions regarding realism. The argument that there are multiple realities is a relativist argument. The related but different argument that there are no theoretical entities is an antirealist argument. He says that the debates here are in fact realist *versus* antirealist, not realist *versus* relativist. Phillips finally asks: if observation is theory laden and if we have no foundationalism, does this not strip objectivity of any meaning? Here he takes the general postpositivist position that objectivity is a regulative ideal (1990b). Thus "the objectivity of an inquiry does not guarantee its truth . . . *nothing* can guarantee that we have reached the truth" (1990a, 43). Speaking carefully, we should note that it does not follow that foundationalism leads to strong correspondence. There may be foundations in the sense of incorrigible truths or experiences, but the superstructure may nonetheless be difficult to "build." Having said as much, we may well concur with Raymond Williams when he says "it does not end but only begins a controversy . . . when it is said that the purpose is 'to show things as they really are' " (Williams 1983, 259).

Causality

Having sketched some of the key issues that arise when we talk about science as part of social work, it will be useful to see how these issues influence how we think about causality. A helpful account of mainstream views about causality in social work is given by Palinkas (2014). In summary, his position seems to be that "the path to causality can be viewed as moving across a series of steps that begin with identification and proceed to description, explanation generation, explanation testing, and prescription or control" (544). As for the means of accomplishing this process, "in social work as in other fields, priority in the determination of causality is given to quantitative methods in general and RCTs in particular" (542). He acknowledges ethical and logistical limitations that are, he believes, inherent to relying on RCTs to determine causality, for which qualitative methods will sometimes be advisable. He relates the respective role of quantitative and qualitative methods in terms akin to traditional language regarding pilot studies.

> Qualitative methods would be especially important in the early exploratory stages of scientific inquiry and for providing in-depth understanding of the causal chain and the context in which it exists. Quantitative methods would be especially important in the later confirmatory stages of scientific inquiry and for generalizing findings to other populations in other settings.
>
> (546)

An example of this pattern of inquiry is given by Webber (2014).

Lying behind this approach to causality is an acceptance of criteria for causality that go back to David Hume and John Stuart Mill, with their perceived implications for ideas of temporality, specificity, measurement, probabilistic inferences, and experimental design. Inasmuch as qualitative social work writers say anything about causality, the backcloth image sometimes is the same (e.g., Padgett 2008).

The humanist alternative starts into this issue from the position that "mind" and "matter" are different entities. Different schools have interpreted the distinction in different ways, but the distinction always means that "mind and matter of necessity have to be studied by methods appropriate to their respectively separate realities, neither being reducible . . . to the other" (Hughes 1980, 66). The humanist program was in part a reaction against a scientized conception of the human actor. The issues are encapsulated in Weber's famous characterization of "social action"—"an action is social when a social actor assigns a certain meaning to his or her conduct and, by this meaning, is related to the behaviour of other persons" (quoted in Hughes 1980, 71).

The key question is whether meaning, particularly in the sense of intentionality, is compatible with causal explanation. The answer has consequences for the relation of the natural and social sciences. There is a well-known example of traffic behavior at traffic lights, where the argument goes that while this may appear like causal behavior, the relationship between lights and traffic is a meaningful one, and what we have uncovered in seeing drivers' response to changes in the light color "is a custom or rule governed practice rather than an instance of a causal law" (Hughes 1980, 72). Drivers could give a reason why they stopped—an *account*. One piece of behavior can be part of many different actions. For example, what might lifting an arm mean? To assault, to give a toast, to get a drink, to greet someone, to ask the team car to come to the bike rider in the Tour de France.

Hence descriptions are also always incomplete because selective. Therefore, although often adequate for the purpose, they can always be contested.

The adequacy of the conventional position in causality may be questioned for other reasons. A brief consideration of analytic induction raises important issues in this regard. Analytic induction first surfaced in connection with the Chicago School of sociology. The first reference clearly identifying analytic induction was in Znaniecki's *The Method of Sociology*,[13] first published in 1934, where he developed it as part of his argument against enumerative research and the growing influence of statistical methods. The method is also in contrast to either theoretical speculation or simple description as in most forms of ethnography. The method was adopted in studies of opiate addiction (Lindesmith), financial trust (Cressey), and, perhaps best known in social work, marihuana use (Becker). More recently the method has been employed and written about by Bloor (1978, Bloor and Woods 2006). Gilgun is one of few social work writers to reflect on analytic induction, albeit modified, as part of her arguments for deductive qualitative analysis.[14] Becker's work will serve as an illustration (example 1.2).

EXAMPLE 1.2
Becker and Analytic Induction

"In doing the study, I used the method of analytic induction. I tried to arrive at a general statement of the sequence of changes in individual attitude and experience which always occurred when the individual became willing and able to use marihuana for pleasure, and never occurred or had not been permanently maintained when the person was unwilling to use marihuana for pleasure. The method required that *every* case collected in the research substantiate the hypothesis. If one case is encountered which does not substantiate it, the researcher is required to change the hypothesis to fit the case which has proven his original idea wrong . . .

"an individual will be able to use marihuana for pleasure only when he goes through a process of learning to conceive of it as an object which can be used in this way. No one becomes a user without (1) learning to smoke the drug in a way which will produce real effects; (2) learning to recognize the effects and connect them with drug use (learning, in other words, to get high); and (3) learning to enjoy the sensations he perceives."

Source: Becker (1963, 45, 58).

There are several key elements in Becker's account that bear on how we think about causality in social work science. Expressed generally, this is clearly a much stronger notion of causality than that employed in mainstream thinking. It seeks to identify both necessary and sufficient conditions for providing a causal link. Unlike probabilistic statements arrived at through enumerative and statistical methods, Becker speaks of a requirement that *every* case should substantiate the hypothesis such that if only one case is encountered that does not substantiate it, the researcher is *required* to change the hypothesis. Second, it is a case-based rather than statistical method. In this it shares Ragin's approach in comparative qualitative analysis (e.g., Ragin 1997). Third, it involves following repeated processes in a loop.[15]

Hammersley (2012) summarizes the general process of analytic induction. It starts by specifying an outcome (positive or negative) to be explained. The researcher collects data on a small number of cases in which the outcome is present. She identifies what those cases share and formulates a hypothesis on that basis. She then collects data on a new small number of cases. If one or more cases are found that do not fit, then either the hypothesis must be reformulated or possibly the outcome. Cressey worked this way in developing what eventually became an outcome of "financial trust violation." And so on and so forth until new cases no longer require revision.

Analytic induction—especially the classic rather than modified versions that commence with a hypothesis—is interesting and significant because it raises several important questions for social work science. It "opens up to challenge at least some of the assumptions built into the initial formulation of the research question" (Hammersley 2008, 69), including the conceptualization of the phenomenon to be explained. It also poses questions such as: What is the form of an adequate scientific explanation? What is required of us to develop such an explanation? What kind of causal relations exist in the social world? Under what conditions do those relations operate?

An important question for social work science is the relationship between analytic induction, enumerative induction, and abduction. To express this precisely, it is possible to believe that analytic and enumerative induction methods are complementary and thus essential parts of science while at the same time concluding that only the former is adequate to understand causes. This is because it is "superior to the various philosophical attempts to find a logical or probabilistic rationale for induction; and also to Popper's argument that the source of theoretical ideas is irrelevant, that only the deduction of hypotheses and their testing is part of scientific method" (Hamersley 2008, 85). We pick up on abduction in chapter 7.

What might we conclude? McCrea, in her comprehensive essay, resolves that "researchers need a diversity of causal heuristics from which to choose" (2007, 272). She is addressing what she sees as a core problem: "*It is not just the existence of alternative unmeasured causes that can compromise the validity of research findings, but also that those causes may work in radically different ways from ones the researcher has taken into account.*" Recognizing that causes can interact with each other, work in tandem, be circular, emergent, necessary or sufficient, distal or proximal, and so on, she concludes: "*Evaluating the causal validity of research designs responds to the central question: Is this design as a whole a sound heuristic for studying the causal process(es) that are the focal points for this study?*" (272). To believe that some variant of experimental design is needed in all cases to study causes is at best an oversimplified understanding of causal realities. To say that something determines something else takes us only a small way toward explaining causes.

Kuhn and Paradigmatic Science

"Paradigm," "scientific revolutions," and "normal science" are persistently troublesome words. Kuhn's own multiple uses of terms like "paradigm" and his apparent shift in position over time (cf. Kuhn 1992) do not help. But neither does the tendency, sometimes encountered, to assume these are grounds for failing to understand his work. Example 1.3 allows Kuhn to speak in his own words, along with occasional remarks from myself.

EXAMPLE 1.3

Kuhn on Science, Paradigms, Normal Science, and Revolutions

Preparadigm science. Kuhn allows "there can be a sort of scientific research without paradigms" (11). He describes it as "a condition in which all members practice science but in which their gross product scarcely resembles science at all" (101). He would include much social science when he says, "That pattern is not unfamiliar in a number of creative fields today, nor is it incompatible with significant discovery and invention" (13). "It remains an open question what parts of social science have yet acquired such paradigms at all" (15).

Paradigms. "Paradigms provide scientists not only with a map but also with some of the directions essential for map-making. In learning a paradigm the scientist acquires theory, methods and standards together, usually in an inextricable mixture" (109)—they are constitutive of the research activity and of the science per se. The paradigm transforms groups into a discipline/profession. Proponents of competing paradigms "practice their trades in different worlds. . . . Practicing in different worlds, the two groups of scientists see different things when they look from the same point in the same direction" (150). This is not to say they can see anything they please. "Both are looking at the world, and what they look at has not changed" (150). But they notice different things and in different relations to one another, so that one theory seems intuitively obvious to one but "cannot even be demonstrated" to the other. Scientists' "communiqués" also change, from books addressed to anyone interested to "brief articles addressed only to professional colleagues." He describes these forms of communication as "obvious to all and oppressive to many" (20). He stresses (in what is one of his important points about a paradigm) that it is sharing the overall paradigm that unites, not sharing constituent rules, assumptions, or points of view (that is, there will be important diversity within a paradigm).

Normal science is "sufficiently open-ended to leave all sorts of problems . . . to resolve" (10). Yet he also says, "Normal science does not aim at novelties of fact or theory and, when successful, finds none" (52)—indeed, it may resist those who propose them. A paradigm does not equal a pattern for replication but an object for articulation. Normal science *is* that articulation, through a kind of (fascinating) "mopping up operation" (24). The paradigm enables the solution of problems that would not have been imagined without the paradigm. There is extensive *fact gathering* of three kinds: facts that "the paradigm has shown to be particularly revealing of the nature of things" (25); facts that can be compared with predictions for the paradigm—"the existence of the paradigm sets the problem to be solved" (27); and empirical work to articulate the paradigm theory (that is, determining universal constants and laws).

There are also *theoretical* problems for normal science. There are "immense difficulties often encountered in developing points of contact between a theory and nature" (30). "These three classes of problems—determination of significant facts, matching of facts with theory, and articulation of theory—exhaust . . . the literature of normal science" (34). But this is not *all* of science, as there are also "extraordinary problems" that may create anomalies and lead to:

Scientific revolutions. But even severe anomalies may not lead to paradigm rejection. "The act of judgement that leads scientists to reject a previously accepted theory is always based on more than a comparison of that theory with the world." It rather involves "the comparison of *both* paradigms with nature *and* with each other" (77). Early-paradigm research will follow the rules quite closely, but increasingly the paradigm gets modified even to the stage where it is not agreed what the paradigm is (83). The effects of a crisis are "the blurring of the paradigm and the consequent loosening of the rules for normal research" (84), followed by a change to a new paradigm where "the profession will have changed its views of the field, its methods, and its goals" (85).

Source: Kuhn (1970).

There is much that could be said about Kuhn's essay on science and scientific revolutions, including matters not represented in example 1.3. For example, he has plausibly denied from his earliest writing on this theme that he holds a relativist position. He remains on balance committed to believing that "scientific fact and theory are not categorically separable, except perhaps within a single tradition of normal scientific practice" (Kuhn 1970, 7). The "except" is a very extensive caveat. Hence he does not entirely reject distinctions of description and interpretation, of "context of discovery" and "context of justification": "I still suppose that, appropriately recast, they have something important to tell us" (9). "Discovering a new sort of phenomenon is necessarily a complex event, one which involves recognizing both *that* something is and *what* it is"—"both observation and conceptualization, fact and assimilation to theory" (55)—and so it is a process over time.

Nature itself, whatever that may be, has seemed to have no part in the development of beliefs about it. Talk of evidence, of the rationality of claims about it, and of the truth or probability of those claims has been seen as simply the rhetoric behind which the victorious party cloaks its power. What passes for scientific knowledge becomes, then, simply the belief of the winners.

These words are interesting more for who spoke them than for what they say. Kuhn's name is too often taken in vain for a relativist view of science, whereas for him the position just quoted is "absurd: an example of deconstruction gone mad" (1992, 8, 9). For him observations are not "givens" but "collected with difficulty" (1970, 126), and scientists with different paradigms engage in different observation methods. "It is hard to make nature fit a paradigm. That is why the puzzles of normal science are so challenging and also why measurements without a paradigm so seldom lead to any conclusions at all" (135). In his postscript to the 1970 edition he says he had only recently realized that being persuaded is not the same as conversion. "To translate a theory or worldview into one's own language is not to make it one's own." For that to happen, he says, "one must go native, discover that one is thinking and working in, not simply translating out of, a language that was previously foreign" (204). But that transition is not typically a conscious choice but something that one slips into without realizing.

Where does this take us? The attempt to expunge metaphysics from human affairs has been abandoned, following the work of Popper and Kuhn. This is part of a move away from atomism and empiricism and to a recognition of how meaning is related to social context. Kuhn shared with Popper the view of "science as a collective enterprise, an institutionalization of critical reason" (Giddens 1993, 142). But against Popper, Kuhn suggests that the development of science depends on the suspension of critical reason, through the taken-for-grantedness of a set of epistemological assumptions, in respect of the underlying premises of paradigms, and that scientists often explain away inconsistent findings. Like magical and religious practices in nonindustrial societies, science is able to "explain" nonconfirming instances. While Kuhn has less to say on this, his position takes us almost directly to the question of the relation of science to forms of knowledge that would normally not be regarded as science.

The somewhat provisional nature of aspects of Kuhn's work has allowed critics to misread parts of his argument as standing for the whole. Even Giddens falls into this trap when he alleges that Kuhn exaggerates the unity of paradigms—"the taken-for-granted, unexamined assumptions shared by communities of scientists, who confine their attentions to small-scale puzzle-solving within the bounds of those assumptions" (1993, 149). In reality, Giddens insists, there are deep-rooted differences between rival theoretical schools. A "potential scepticism regarding the claims of science is in a fundamental sense built into the legitimate order of the social organization of science—even if not consistently acted upon" (150). We consider whether there are scientific communities in social work in chapters 2 and 5. But to represent Kuhn as speaking of normal science as small-scale puzzle solving within a set of unexamined assumptions is barely credible from a full reading of Kuhn. Kuhn did, indeed, include a chapter on normal science as puzzle solving, where he argues that normal science does not aim at major or unexpected novelty but does require expert puzzle-solving skills. A "good" puzzle is not good because the outcome is important. A puzzle has an assumed solution. Normal science tends to focus on solvable puzzles. Puzzles also have very binding rules—almost like preconceptions. But it is this exact context, Kuhn goes on to say, as we have seen in example 1.3, that it is sharing with the overall paradigm that unites; it is not just sharing constituent rules, assumptions, or points of view.

Giddens is on safer ground when he says that Kuhn does not do justice to how "the development of science is constantly interwoven with, and affected by, social interests and influences that nominally stand outside of science itself." The "institutional autonomy of science is . . . never more than partial" (150).[16]

Social Constructionism

Social constructionism and hermeneutics figure frequently in the following pages. Here my only reason for including a brief section is to set out a few central terms. A strong *social constructionist* would respond to both postpositivists and critical researchers by saying:

1. Facts are social constructions rather than objects.
2. It is constitutive of a given fact that it is so constructed.
3. The construction is contingent and not universal.

The interest lies in exposing constructions where none is thought to exist—"where something constitutively social had come to masquerade as natural" (Boghossian 2006, 18). To "de-construct" is then thought to be potentially liberating. This is a strong version of social constructionism. Schwandt expresses his own species of constructivism as involving the positions

> (1) . . . that the social world . . . can only be studied from a position of involvement "within" it, instead of as an "outsider"; (2) that knowledge of that world is practical-moral knowledge and does not depend upon justification or proof for its practical efficacy; (3) that we are not in an "ownership" relation to such knowledge but we embody it as part of who and what we are.
>
> (1997, 75)

The classical philosophical view is that it is sometimes possible for the evidence alone to explain why we come to believe something. Task-centered social work may be taken as a hypothetical case in point. The original experiment, *Brief and Extended Casework* (Reid and Shyne 1969) was carried out with the assumption that open-ended casework would prove more effective, given ideal circumstances. When the contrary was supported by the evidence, it led to a prolonged program of research to develop and test the effectiveness of the model. A strong constructivist, as described above by Boghossian, would respond by asserting the *descriptive dependence of the facts* in this experiment and would be deeply skeptical of evidential truth claims based on Reid and Shyne's experimental data.[17] A moderate constructivist would respond by asserting the weaker position of the *social relativity of descriptions*. He or she would accept that evidence may exist independent of description and language but hold that we cannot assert that something is true, or that our warrant is unchallengeable, or that it will forever be warranted. Blumer, for example, argued that a constructivist view

> does not shift "reality," as so many conclude, from the empirical world to the realm of imagery and conception . . . [The] empirical world can "talk back" to our picture of it or assertions about it—talk back in the sense of challenging and resisting, or not bending to, our images or conceptions of it. This resistance gives the empirical world an obdurate character that is the mark of reality.
>
> (1969, 22)

We saw earlier that even Guba came out with a surprisingly realist phrase regarding "nature's propensity to confound" (1990, 19). Not all constructivism is incompatible with realism. On this view of things we can maintain belief in the existence of phenomena independent of our claims about them, and in their knowability, "without assuming that we can know with certainty whether our knowledge of them is valid or invalid" (Hammersley 1992, 50). Hence, even if—like me—you go for something more toward the weaker position, judgment about evidence still remains a hugely demanding task both methodologically and, as I discuss earlier, philosophically. Because we accept that there are some mind-independent facts, "This argument . . . does not tell us all by itself which facts obtain and which ones don't; nor does it tell us, of the facts that do obtain, which ones are mind-independent and which ones aren't" (Boghossian 2006, 57). Interestingly, Bill Reid seemed to reach a position something like this when, in a chastening and memorable metaphor, he remarked of effectiveness studies, his own and those of others: "It is like trying to decide which horse won a race viewed at a bad angle from the grandstand during a cloudburst" (Reid 1988, 48).

One more caution is in order. Some social work writing implicitly links positivism with the natural sciences and constructivism with social work science. This is inadequate. Even science of the natural world can be understood as socially constructed. We return fairly extensively to challenges posed by constructionist science in chapters 7 and 8. However, for the humanist alternative, "the difficulties centred on the nature of understanding and the criteria of adequate understanding, social and cultural relativism, and the relationship between actors' concepts and those of an observer" (Hughes 1980, 123).

But having spent this opening chapter speaking less explicitly about social work than may have been anticipated, in the next chapter we turn from *talking about* to *doing* social work science, and its relation to other forms of knowledge.

Taking It Further

The annotated readings may be taken before or following the tasks, here and throughout the book.

Reading

Howe, D. 2014. *The Compleat Social Worker.* London: Palgrave Macmillan.

I was not able to consult this before completing this book, but David Howe always writes basic texts in ways that invite the reader to think afresh about familiar questions. As the publisher's note says, "*The Compleat Social Worker* explores the many debates the profession enjoys, including those between nature and nurture, care and control, thought and feeling, art and science, facts and values. In examining these ideas and the discussions they sponsor, it celebrates social work's rich heritage of scientific thought and human relationships."

Phillips, D. C. 2000. *The Expanded Social Scientist's Bestiary.* Lanham, Md.: Rowan and Littlefield.
Williams, R. 1983. *Keywords: A Vocabulary of Culture and Society.* London: Fontana.

Williams's book has been, to use a forgotten expression, my *vade mecum* for many a year. Whichever edition you locate, it will reward with constant surprises, joys and insights. Phillips is insightful and helpful to read alongside Williams.

Task

Take two or three contrasting research articles by social work writers. Working with someone else, identify what positions they each take in their work on:

- The place of theory
- If and how causality can be understood
- The relationship of their findings to the "real world"

As you work on this task, make a general note on how science language is used by each author.

Positions may be implied rather than explicit, but your judgments should be ones that the writers, were they to see them, would regard as a fair expression of where they stand.

[2]

Doing Social Work Science

To speak of doing social work science raises "boundary" questions as to the relationship between science and other forms of knowledge and action. I preface sketches of some key examples with a brief reference to ideas of "action." Personal knowledge (for example, experience and commonsense), art, political action and power, faith, and ideas of skill fill out this part of the chapter.

I move on to think about doing social work science and, finally, doing *good* social work science.

To speak of the first chapter as "*talking about* science" and this chapter as "*doing* social work science" points to but of course overstates a distinction. In opening questions of theory and its relation to practice we have already stepped into the doing of science. And in the substantial part of this chapter on the relation of science to other forms of knowledge—forms that often are more associated with doing—we hark back implicitly to questions of how we should understand the nature of science.

Kinds of Knowledge

EXAMPLE 2.1
A Narrative of Pain

I watch her being wheeled into the room on a stretcher. I am certain she can walk. She is organized onto the table—laid out as a specimen. Everyone else settles into their chairs, as if preparing to watch a film at a neighborhood cinema. We begin, differently. The technician types on the keyboard. "Any allergies?" "No, none." I sit in front of a panel that remotely controls her. Two screens display images of "her" simultaneously. I see her through the glass, lying still—not looking like these images at all. Motionless. What is she thinking? What is she afraid of? Is she afraid? The control begins—she is moved forward, backward, back a little more. The technology, in particular the movement, defines her insofar as she literally becomes embedded in this practice. She moves away from me. She is moved away. In the ambulance on the way here for the CAT scan she tells me that the last two months without radiation has been the first time her hair has grown back in what seems like years. She describes her reflection in the mirror as if it is not her. She shows me, in the polished parts of the ambulance doors, how she sees herself. This sickness has become her.

I shift my eyes to the computer screen to my left, then to the next screen, and I see circles of white appear as rapid images of her brain flash on the screens one after the other. I stare at these white circles as they become better defined. The breast cancer has spread to her brain. Images that have meaning. Real meaning. The technicians share a joke. I have learned that the more they joke the worse the prognosis. She is loaded with brain metastasis. I want to shout. You are dying. We are all dying, but you are going faster than us.

Source: Phillips (2007, 201–202).

I am never unmoved when rereading this article by Catherine Phillips. She says she sketches a narrative from her own clinical practice as a social worker in an emergency department in an acute-care hospital in Toronto. The narrative is drawn from multiple interactions.

> It has been assembled . . . to demonstrate two things: one, that clinical interactions are a series of social texts; and two, when studied closely, such interactions are replete with everyday acts of power that must be

> attended to in social work research into clinical practices. . . . It is my hope that the writings of experience in this article can confront the fictions, fantasies, narratives, explanations, and signs that allow patients in pain a limited number of transgressions.
>
> (Phillips 2007, 201)

It opens this chapter as a way of jolting us out of thinking only "scientifically" about the kinds of things we may know.

Phillips says of a later moment:

> I talk to her on the phone at various junctures, ensuring that the community-based services are in place. She tells me that a nurse came to see her today and showed her pictures of faces and asked her to indicate which face described her pain the best—smiling face, taut face, stunned face, laughing face, grimacing face, unhappy face. Unintelligible faces. She tells me that she cannot differentiate her physical pain from her pain of dying. In this surprise and desire for life is an expression of pain that defines her as a subject. Stoned with pain, she is silent on the phone. She becomes a subject for merely a moment through these interpretations of pain, and then seamlessly returns to being an object.
>
> (202)

Expertise, professional skills, power, technology, meaning—these all lie not far below the surface. And none without some element of ambiguity, reminding us again of the opening words of this book as to how science and the forms of knowledge that border it are an inescapable presence in social work. In what ways can or should we distinguish social work science from other kinds of knowledge? We can distinguish views on this question in a very simple way according to the extent to which people think "science" knowledge is *similar to* or *interwoven with* other kinds of knowledge, such as knowledge from experience (for example, "practice wisdom"), common sense, tradition, authority-based knowledge, faith, magic, and so on. It may help to separate out the elements of this question by distinguishing the relationship between:

- Conscious, self-aware knowledge and tacit, intuitive, practice-wisdom knowledge
- Rational, scientific knowledge and authority-based, traditional knowledge

- Expert knowledge and experience, common sense
- Theoretical knowledge and practical knowledge

The famous essay by C. P. Snow (1956) on "the two cultures" still serves as a reference point for such discussions. He called these the "traditional" (mainly literary) and the "scientific" and lamented the "precious little communication" between them. The scientific culture is expansive, "certain that history is on its side, impatient, intolerant, creative rather than critical, good-natured and brash." He opines, "neither culture knows the virtues of the other; often it seems they deliberately do not want to know." Snow wittily observed that "the difference in social manners between . . . Los Alamos and Greenwich Village would make an anthropologist blink." For example, "sometimes it seems that scientists relish speaking the truth, especially when it is unpleasant." Snow asked what the loss was for those within the traditional culture. "Those without any scientific understanding miss a whole body of experience: they are rather like the tone deaf, from whom all musical experience is cut off and who have to get on without it."

Art, practice, the practical, experience, expertise, common sense, faith, tacit knowledge, lay knowledge, personal taste, and even skill or technology are some of the words and phrases that are placed in counterpoint to science. In terms of how this relates to science and social work, two observations are in order. First, a simple binary categorization of the distance between science and other forms of knowledge is almost always the wrong place to start. Second, the level and extent of social work attention to these issues has on balance been inadequate and deleterious in its consequences for the field. There are exceptions to this charge of neglect around, for example, expertise, art, and practitioner forms of knowledge, but some of the more interesting earlier contributions do suffer from later neglect.[1]

A major source of lost opportunity in this area lies in the lack of "translation work" of key writers in other disciplines and the failure to recognize the potential relevance of certain fields. For example, an area of thought that does not seem to have been addressed in social work is the question of what differences should or can be regarded as differences of *judgment* and what should or can be seen as matters of *taste*. The distinction rests in seeing matters of judgment as "discussable" such that evidence and argument can be brought to bear. "Taste" is "incorrigibly subjective; it is private; and for that reason there is nothing that can be said about it that has any consequence" (Shapin 2012, 172). In this connection we have little empirical understanding of questions such as: How does one become an enthusiast for, for example, randomized control trials? How does one become a good researcher?

To hint at other possibilities, the generally thin discussion of matters of faith in relation to social work science would be enriched and unsettled by considering how arguments for objectivity have been dealt with by some reflecting on Christian history (e.g., Hooykaas 1972, McKay 1987). Again, the rather undifferentiated ways in which "practice wisdom," common sense, and experience-based knowledge (as in the phrase "experts by experience") are too often treated would be diversified by looking at what some have done around the notion of disinterestedness among citizens (e.g., Evans and Plows 2007, Schutz 1967b) and connoisseurship (Eisner 1988, 1991; Polanyi 1958). To suggest the possibilities of "translation work" I include here an account of a research study that will seem as far from social work as one can go—an article that starts with the story of a dam in Idaho that failed in 1975—but because of its contextual strangeness helpfully gets us to consider kinds of knowledge and their relationship to each other (Schmidt 1993).

Grout

The case came to a committee in Congress. The grouting expert called to testify explained that grouting is not an exact science and that absolute certainty is not possible. Grouting, he explained, is more like an art and requires a certain "feel" for the work. To save wastage of grout being washed away through the cracks, salt was added to speed up setting time. If the holes were larger, then sand also was added. But no formula could be found to relate the amount of salt to setting time for the grout. So "the decision on whether to add salt or sand, and how much of each, was left to the grouters and their mysterious 'feel for the hole'" (Schmidt 1993, 526). This knowledge is difficult to put into words, as it depended on judgments about the resistance of rock to drilling; the color, pressure, and even the smell of the water flowing back; the humidity of the air; and so on. So judgment could never be separated from the context nor subjected to analysis. The grouters "must be constantly alert to the 'talk-back' of the specific situation" (526).

There were other workers who operated at the top of the dam. On one occasion they had received instructions to stop work. They were bewildered.

They tried to construct a plausible explanation and thought that, though the rocks were equally fractured at the top as at the bottom, the water pressure was less at the top so needed less grouting. "These workers demonstrated a kind of passive—yet critical—knowledge, as when one understands a language but cannot speak it, or appreciates a good design but cannot create it." As Schmidt expresses it, the grouters had "a feel for the hole" whereas these workers had "a feel for the whole."

There were others with other forms of passive yet critical knowledge, too. Forest rangers often monitor the dams on their territory and develop an intimate knowledge of them over time. People who regularly fish in the area or hike may also gain such knowledge. They may detect problems but may assume that people in charge already know about the problem. They might also worry that they would risk humiliation by speaking out. So here we have a feel for the unknown (the grouters), passive/critical knowledge, and intimate knowledge by experience.

Why were these different kinds of knowledge ignored? In part no doubt because of power and status among scientists, engineers, and managers. But also ordinary people often consider many technologies with a sense of "dread." Schmidt helpfully suggests that these different kinds of knowledge have shared general characteristics:

1. They usually focus on specific phenomena as opposed to abstract classes of knowledge.
2. They typically need direct, bodily involvement in acquiring such knowledge rather than through impersonal instruments.
3. There is a need for the synthesis of data from several senses, in contrast with the analytical approach of science.
4. They are characterized by the qualitative nature of knowledge and are not reducible to quantitative terms.
5. They are often difficult to describe in words, entailing, in Schön's words, that "we know more than we can say."[2]

Compare this example of a Japanese social worker talking about practice knowledge and learning:

> Looking back on my learning experiences, I found myself learning a lot when I reflected on what would be the causes of a case which went well or wrong. If anything, I found myself learning a lot when things

> went wrong. In such occasions when I felt impatient as a professional, I sometimes attended a training course and workshop which was of interest to me, and then I sometimes found something I learned in those opportunities fitting into what I sought for. I mean, for example, a lecturer explained the things that I had not been able to put into words well. Or, I found out in a book the things that I had wanted to express. I suppose, I have repeatedly been learning something in those ways as a professional.[3]

Here we have reflection, retrospection, a process of making a judgment about "fitting" together different learning sources (books) and contexts (training courses and case interventions), putting "things" into "words," and judgments about things going wrong.

Action

Schutz speaks of action as "human conduct devised by the actor in advance, that is, conduct based upon a preconceived project" (1967a, 19). "Action may be covert (for example, the attempt to solve a scientific problem mentally) or overt, gearing into the outer world; it may take place by commission or omission, purposive abstention from acting being considered an action in itself" (20). In this connection he introduces the idea of motives and distinguishes:

> 1. Motive as the state of affairs that is to be brought about by the action. He calls this the "in-order-to-be" motive.
>
> 2. The point of view of the actor to those past experiences of his that have determined him to act the way he did. He calls this "(genuine) because-motives."
>
> (22)

He regards this distinction as "of vital importance for the analysis of human interaction" (22). What of the observer of actors? The observer is "tuned in" on the motives of the actors, but they are not tuned in on his. "He is not involved in the actor's hopes and fears." It is this that constitutes his "disinterestedness." But "it is merely the manifested fragments of the actions of *both* partners that are accessible to his observation." "There is a mere chance, although a chance sufficient for many practical purposes, that the observer in daily life can grasp the subjective meaning of the actor's acts" (26). Thinking of the social work scientist, "the scientific observer of human inter-relational patterns . . . has to develop specific methods for the building of his constructs," in particular "constructs of models of so-called rational actions" (27).

If we think in the ways suggested by Schutz, we ought to be at least as much interested in what practice-action has to say to science as vice versa. "Knowing" and "doing," science and practice, are not two wholly distinct areas that need mechanisms to connect them but are to a significant degree part and parcel of each other. Two of the most stubborn and difficult-to-avoid options for presenting the relation of science and practice are to give science priority over practice (rationalism) or to give practice priority over science (romantic conservatism).

In the first kind of case social workers often face the dominance of social science "experts" over practice "beneficiaries." This tends to lead to a deeply unhelpful situation in which practitioners are routinely blamed for their perceived failure to act on the "findings" of science. Kirk and Reid (2002) elegantly criticized "practitioner-blame" responses that take the form of discussions of science as progress, of science as having to struggle against "organizational banality," and practitioners as subverting research and easily being threatened.

A particularly helpful way in to understanding these issues was provided by the systems theorist Norma Romm. Her basic premise is that "the process of attempting to 'know' about the social world already is an intervention in that world which may come to shape its constitution" (1995, 137). The issues can helpfully be unlocked through the recognition that science and practice need linking in ways that release the potential for practice to challenge social work science and in so doing contest conventional hierarchical ways of seeing expert/beneficiary relationships.

I have tried to do this in different ways. First, through research on what is happening when social workers engage in research while in their role as practitioners (e.g., Shaw and Lunt 2011, 2012). Second, through developing a model of qualitative social work that argues for a methodological practice (e.g., Shaw 2011a). Third, through trying to think through the relationship between what we might call inner science criteria for judging research (rigor, validity, and so on) and outer science criteria (relevance, emancipation, utility) (Shaw and Norton 2007). Each of these is based on assuming that a straightforward "application" of research to "practice"—and of "science" to "society" more broadly—is too likely to reinforce a rationalist status quo.

Recall again the original argument for critical theory by Horkheimer back in 1937. The purpose of critical thinking, so he insisted, is not the better functioning of society. "On the contrary, it is suspicious of the very categories of better, useful, appropriate, productive, and valuable, as these are understood in the present order" (Horkheimer 2002, 207). He was concerned that critical thought should not "remain locked up within itself, as happened to idealist philosophy," or engage in "the formalist fighting of sham battles" (211). Despite the proletariat's awareness of contradiction, even that is "no guarantee of correct knowledge" (213). Hence it is to fall short of what is needed when "the intellectual is satisfied to proclaim with reverent admiration the creative strength of the proletariat and finds satisfaction in . . . canonizing it" (214). There is an "ever present possibility of tension between the theoretician and the class which he is thinking to serve" (215).

Personal Knowledge and Common Sense

Social work science is always diminished when its advocates seem to deny—by assertion or more frequently by simple silence—the connections of scientific and personal commonsense knowledge. Polanyi, known for his work on some forms of tacit knowledge, does not strongly separate practical from theoretical knowledge—knowing what and knowing how, *wissen* and *können*. "These two aspects of knowing have a similar structure and neither is ever present without the other" (Polanyi 1966, 20), so he uses "knowing" to refer to both. He refers to lesson-learning experiments in psychology where something is learned but the person is not able to say what has been learned nor how it has successfully been learned. He adopts the expression "attending *from*" (as against attending *to*) the immediate proximal knowing to the distal knowing. The first is tacit, the second explicit. "In general, an explicit integration cannot replace its tacit counterpart." As he pointedly conveys it, "the knowledge I have of my own body differs altogether from the knowledge of its physiology" (20).

He infers that if science seeks to eliminate all personal elements of knowledge, "the ideal of exact science would turn out to be fundamentally misleading and possibly a source of devastating fallacies." "The process of formalizing all knowledge to the exclusion of any tacit knowledge is self-defeating" (20). He asks how we as scientists come to see something as a problem, replying that "to see a problem is to see something that is hidden. It is to have an intimation of the coherence of hitherto not comprehended particulars" (21). Scientific knowledge "is the knowledge of an approaching discovery," even if "the act of discovery, like discovery itself, may turn out to be a delusion" (25).[4]

This raises the question of the relationship between science and common sense and the issue of what Cicourel calls "the necessary and ubiquitous role of commonsense reasoning in science, applied science, and all 'rational' bureaucratic activities" (1985, 172). Writers as far apart on some issues as Alfred Schutz and Donald T. Campbell insist on the centrality of this relationship for understanding the work of science. Campbell—too often colonized by randomized-control enthusiasts who bypass much of his sophisticated work—set out a clear position and in doing so acknowledged the influence of Howard Becker and Erik Erikson.

EXAMPLE 2.2

Campbell and Schutz on Science and Common Sense

Campbell: "Too often quantitative scientists, under the influence of missionaries from logical positivism, presume that in true science, quantitative knowing replaces qualitative, common-sense knowing. The situation is in fact quite different. Rather, science depends on qualitative, common-sense knowing even though at best it goes beyond it. Science in the end contradicts some items of common sense, but it only does so by trusting the great bulk of the rest of common-sense knowledge. Such revision of common sense by science is akin to the revision of common sense by common sense which, paradoxically, can only be done by trusting more common sense.

"This is not to say that such common-sense naturalistic observation is objective, dependable, or unbiased. But it is all that we have. It is the only route to knowledge—noisy, fallible and biased though it be. . . . Even when we improve on it, we must go through it and build on it" (Campbell 1979, 50–51, 54, 60).

Schutz: Schutz places arguments regarding the relationships between personal knowledge and science centrally in every part of the work of science. He nicely expresses it as: "Common-sense knowledge is the unquestioned but always questionable background within which inquiry starts and within which alone it can be carried out" (1967e, 57). "It is the social matrix within which unclarified situations emerge which have to be transformed by the process of inquiry into warranted assertability" (57). On the relation of common sense and science he says, "the thought objects constructed by the social scientists refer to and are founded upon the thought objects constructed by the common-sense thought of man living his everyday life among his fellow-men" (1967a, 6).

Schutz's arguments are important, and I explicate his position in some detail in the following paragraphs. He refers in this context to "second-degree" (sometimes second-order) constructs—"constructs of the constructs made by the actors on the social scene" (1967e, 59). For him this does not mean that social sciences are entirely different from natural sciences, in that "certain procedural rules relating to correct thought organization are common to all empirical sciences" (1967a, 6). It is in this way that he develops his understanding of constructs of thought objects in commonsense thinking, that is, the way in which "the wide-awake grown-up man looks at the inter-subjective world of daily life" (7), or, as he more explicitly says elsewhere when speaking of multiple realities, "a plane of consciousness of highest tension originating in an attitude of full attention to life and its requirements" (1967d, 213).

He takes the position that "the thought objects of the social sciences have to remain consistent with the thought objects of common-sense, formed by man in everyday life in order to come to terms with social reality" (1967a, 43). He concludes, as he expresses it, that the relationship of the social scientist and the puppet he has created reflects the age-old problem of the relationship between God and his creatures. "The puppet exists and acts merely by the grace of the scientist. . . . Nevertheless it is supposed to act as if it were not determined but could determine itself" (44). This takes us to a position similar to Giddens's (1993) argument about the double hermeneutic, which we will return to in chapter 7 to explain how the thought objects in the natural and social sciences are different. The world of nature as studied by the natural scientist does not mean anything to what they study, whereas the world studied by the social scientist means something to those studied. Hence the thought objects of the social scientist have to be founded on those meanings, those thought objects, constructed by the commonsense thinking of people living their daily lives.

Schutz's work is important and of considerable value for social work, and we will return to it later in the book. On how the social scientist proceeds he outlines across several of his essays a process in which we first observe and then construct typical course-of-action patterns, coordinate these with models of an ideal actor, and ascribe purposes and goals (systems of relevances). Each step can be verified by empirical observation, "provided that we do not restrict this term to sensory perception of objects and events in the outer world but include the experiential forms" (1967e, 65).

There is a difficulty here, one Schutz himself is well known for expressing: "How is it possible to form objective concepts and an objectively verifiable theory of subjective meaning structures?" (62). He seems to argue that first-order constructs are "subjective" but second-order constructs are "objective ideal typical constructs" (63). As such they are "of a different kind from those developed on the first level of common-sense thinking *which they have to supersede*" (63; italics added). He is not saying they are objective in the sense of having direct reference to real objects but that they are types.

The difficulty for Schutz (and Weber) is this reference to being able to move between "theoretical social scientist" and "actor on the social scene" (63),[5] which Schutz seems to refer to in two contexts: first, that by becoming a social scientist one moves from one to the other. "By making up his mind to become a scientist, the social scientist has replaced his personal, biographical situation by . . . a scientific situation" (63; cf. Schutz 1967f); second, that in any given situation one may move between one and the other. It is this, of course, that is the basis for how he and Weber regarded objectivity—"detachment from value patterns which govern or might govern the behaviour of the actors on the social scene" (63). Schutz has been criticized by some subsequent writers on the grounds that with such a model "we can only explore how the world can *appear* objective to an individual subjectivity"—it can never distinguish between mere opinion and valid knowledge (Packer 2011, 10).

Art, Creativity, Skill, and Expertise

It is not new to think of social work as art. Timms (1968) left us one of the most penetrating discussions of the question, one that long predates the recent interweaving of the humanities and the social sciences, and England (1986) devoted a whole book to the subject. Adrienne Chambon helpfully points out an article by Norman Denzin (2002), written expressly to a social work audience, where he spoke of bringing social work closer to the arts via a "poetics of social work." She also foregrounds the memory of Howard Goldstein, who was "the most adamant social work scholar" who "repeatedly clamoured for an explicit convergence between social work, the humanities, and the arts" (Chambon 2008, 592).

There are, however, moments when advocacy of the arts feels over the top. Laurie Taylor, the UK media sociologist, gently mocked a well-known qualitative research conference in his "The Poppletonian" column in *Times Higher Education* (August 19, 2010), when hearing of a conference that would "showcase such aspects of 'performative social science' as open-mike poetry, quilt making and circus acts." It recalled for me sitting in a session on hat making in an earlier conference. But *gently* mocking—for Chambon is surely right to mention an exhibit of stitching wedding dresses in rural Canada, as a way of telling through material means of a woman's world and a mother-daughter relationship. She urges through examples how the artist provides "a way of grasping at something not yet named. A 'skin sense' of uncertainties at the edge of an experience that is still unworded" (595).[6]

After negotiating the somewhat labyrinthine problems of scientific and commonsense knowledge, it may seem possible to breathe a sigh of relief in thinking that if any opposites are in general beyond dispute it is that between art and science—between "two cultures," to borrow C. P. Snow once more. An example of a fairly standard view can be seen in McIvor as he talks about this general area. Example 2.3 extracts his key points, expressing them in tabular form.

McIver is quite straight: "the relation of sociology to social work is that of a science to an art." He goes on to aver that "if we fail to recognize the significant difference . . . we shall cherish false hopes and refuse true aids, whether as scientists or artists" (1931, 1). Social science for the social worker "helps us to clear our eyes, to see things steadily and whole, to interpret situations *as though* we lacked the emotions which make us want to interpret them" (4), and it enables the social worker to "advocate further goals

EXAMPLE 2.3
Robert McIver on Science and Art

ART	SCIENCE
"manipulates, controls, and changes"	"seeks only to understand"
"individualizes"	"generalizes"
"lives in its concrete embodiments whether it be sculptured stone or the changed conduct of human beings"	"lives in abstract relationships which it discovers irradiating the concrete world"

Source: McIver (1931, 2).

while still doing the day's work" by giving "a background of intellectual convictions" (7).

Without wishing to discard wholesale what McIver says, we should destabilize his assurance that the difference is so cut and dried. Example 2.4 captures the voices of physicists, chemists, biologists, and other scientists as they talk about their work.

EXAMPLE 2.4
Scientists Talk Art

Carl Djerassi, analytical chemist: "A distinguished scientist who is also successful in the arts" (9). Asked if he saw the creative process in arts and science as different, he says yes, in some respects, in that in science one is aiming to be short and succinct. "So scientific papers are a very ephemeral form of literature. People very rarely reread them. But in poetry . . . the ultimate compliment is to have someone re-read your poems, to remember that book, to remember some metaphor, some nuance that would count for nothing in science" (11). But in the actual creative act he says there *is* similarity. For example, "in both you're doing what hasn't been done before," or so "you flatter yourself" (12). Also in both you gain pleasure.

Roald Hoffmann, theoretical chemist: "In this language simple words like 'power,' 'energy,' 'force,' 'stable,' 'unstable,' acquire a host of alternative meanings. That's partly what poetry is about, about ambiguity, about alternative meanings" (24).

Antonio Garcia Bellido, developmental biologist: On measuring success in science: "Oh, it's very difficult to measure. The one thing that really drives the scientist is the enjoyment of discovery. I mean the pleasure of finding something that explains the phenomena, that may have a universal value. That is, it may go well beyond the understanding of the particular experiment. This feeling is a feeling similar in many respects to the feeling of a creative artist who has finished a painting or a musical composition. It's a feeling of having been in agreement with somebody, with nature in this particular case, in resonance. You have grasped something which is hidden from the rest of the world" (118).

Elwyn Simons, paleontologist: "I think that the best collectors are like great pianists, or great painters. . . . It's a form of seeing. It's seeing order in a random background" (152).

Carlo Rubbia, particle physicist: "We are essentially driven not by . . . the success, but by a sort of passion, namely the desire of understanding better, to possess, if you like, a bigger part of the truth. Hence science, for me, is very close to art. . . . In my view scientific discovery is an irrational act . . . and I see no difference between a scientist developing a marvellous discovery and an artist making a painting" (197). In a passage about how with very small things it entails a philosophical revolution, he says, "there is no *image*, so to speak, of these things that are infinitely small. . . . So it's a genuine trip you're taking inside matter, inside yourself, inside the very objects that surround us" (198–199).

Source: Wolpert and Richards (1997).

The issues lie in part in the way that "art" is what Popper might have called a "bucket word." One need only follow annual national major awards for "art" to realize the elasticity of the term. The word could originally refer to almost any skilled person, and we forget how late the idea that all sciences are arts faded, though it carries a sense that is present in our use of "artisan," a word that came to mean skilled manual worker from the late nineteenth century and that has enjoyed a recent revival.

It was from the same nineteenth-century period that "art" became associated with "creative" and also with expressions like "artistic temperament" (Williams 1983). A check of the use of "art" in social work literature will show that with some exceptions "references to social work as 'art' usually mean social work as skill" (Timms 1968, 74). In this sense social work science and art are not too far distant from each other, recalling the image of science as a golem and "the creature of our art and our craft" (Collins and Pinch 1988, 2).

Art, craft, and skill gradually circle back to personal knowledge. Take the words "expert" and "expertise"—terms we examine in some detail in subsequent chapters. We see the term over against "lay" (in the sense of "of the people"), in the context of the growth of specialization and credentialism. But the word is closely related to "experience"—a word we use both to refer to "knowledge garnered from past events" and "a particular kind of consciousness" (Williams 1983, 126), and thence to "empirical" and "experiment."

There has been a quite influential UK attempt to distinguish yet simultaneously suggest a framework to relate kinds of social care knowledge. Led by the realist scholar Ray Pawson, distinctions were made between types and quality of knowledge in social care (Pawson et al. 2003). They distinguish five kinds of knowledge—organizational knowledge, practitioner knowledge, user knowledge, research knowledge, and policy community knowledge—and provide a quality framework that entails seven quality dimensions:

1. Transparency: is it open to scrutiny?
2. Accuracy: is it well grounded?
3. Purposivity: is it fit for purpose?
4. Utility: is it fit for use?
5. Propriety: is it legal and ethical?
6. Accessibility: is it intelligible?
7. Specificity: does it meet source-specific standards?

The contribution this work makes is not so much to the development of quality criteria—these seven dimensions are similar to criteria elsewhere in the literature—but in the argument that it may be feasible to think of developing a set of criteria that is generic to very different kinds of knowledge. While this is useful—it prevents us, for example, from setting up hierarchies of knowledge and/or neglecting the numerous people who have something to bring to the discussion—I think it assumes too much about the comparability of judgments to assess different kinds of knowledge, and it also seems to assume there is a generic (for example) practitioner knowledge rather than knowledge (of different kinds) that practitioners possess.

Faith and Social Work Science

Faith-based stances on social work in general and social work science in particular have perhaps been less self-effacing in the United States than in most countries in Europe and elsewhere. It was the English poet Matthew Arnold who, in his poem "Dover Beach," long ago remarked on "the melancholy, long, withdrawing roar" of the "Sea of Faith" retreating. My reason for including the following discussion is to suggest illustratively rather than comprehensively the interest of two examples: Islam and a form of Christianity that flows out of the sixteenth-century Reformation.

One reason for choosing them is that there are various obvious reasons why these may be thought of as among the least likely candidates for positive press on their contribution to social work science.[7] In the case of Christianity, this lack of expectation flows from how social work science and the social sciences in general are the heirs of skepticism about creedal forms of Christianity. The archive at the University of Chicago illustrates this. Take as an instance Albion Small, the first head of department for sociology. There is an interesting note in the archives dated March 19, 1924, that Small heads "My Religion."[8]

> My Religion: Is my attempt to make Jesus Christ the Pattern and Power of my life:—
>
> It is my attempt by all means at my command to find out what the Pattern and Power of Jesus Christ mean in terms of my own daily work:—
>
> It is my attempt to frustrate the tendency of my theology to displace my religion.

He says: "When I want to distinguish this religion from any and all predominantly intellectual schemes of belief, I call it *Christianism.*" He says this is his "quota towards solution of the problem of forming a solid front against Fundamentalist theologizing, and of symbolizing it." In an earlier article Small favorably quoted a "Christian minister" as saying:

> There must be clearer ideas of the fatherhood of God, and the brotherhood of man; there must be reconstruction of ideas concerning the independence of the individual as related to the solidarity of society; there must be revision of our ideas of the sacred and the secular; there must be reconstruction of our ideas of property; we must clarify our views of the relations of religion and politics; we need to reconsider the relation of individual to public opinion; we need to detect modern Pharisaism; we need to overcome irrational partisanship.
>
> (Small 1896, 568)

Barnhardt, a student during Small's tenure, later recalled former ministers who were students with him and remembered Small saying, "It looks as though I'm going to be able to get all these preachers to study sociology—that's a good thing" (Interviews with Graduate Students, Box 1, Folder 1).[9] We will see more of this in chapter 3 in relation to the religion of science, but Carey argues that by the 1920s radical social Christianity of this kind was faltering and being replaced by a concern with how to apply sociological knowledge (1975, 65). George Herbert Mead (1923) was interested in understanding this process. He took the position that "the intelligible order of the world implies a moral or social order, i.e. a world as it should be and may be" (245). He takes this back to Augustine and Milton of a plan of salvation. Although "the sharpness of the Plan has faded" following the advent of natural science, "the idea that the universe is in some way geared to the intelligence and excellence of our asocial and moral order has not disappeared from the back of men's minds" (230). But this had become "profoundly different" from its Christian origin perhaps because of "the insistent curiosity of recent science, which has refused to accept any given order of nature as final" (231). The teleological and spiritual had been replaced by the mechanical and materialistic. Interestingly there is a short note on "Methodology of Social Sciences" in his papers in the University of Chicago archives, which he laid aside and finished midsentence, where he says "the most fundamental social science is that of Ethics. . . . The other social sciences are all subsidiary to that of Ethics. . . . Sociology seems to me to deal more specifically with appearance and change of societies and their institutions in terms of the conduct of individuals and groups" (George Herbert Mead Papers, Box 13, Folder 20).

Islam and Social Work Science

Most general accounts of social work and science in Muslim societies are liable to be countered. Yet we can perhaps accept as a generalization that within Arab Muslim society religious knowledge is both the paradigmatic and most highly valued form of knowledge. Within a traditional Arab culture the key issue is the culturally perceived relationship between a given field of study and the religious sciences (that is, those bodies of knowledge that are thought to elucidate the Koran). Lines of continuity rather than discontinuity are sought from "old" to "new" knowledge, so that social work science may tend to be seen as old religious philosophy writ large.

This is not to say that there is no scope for "secular knowledge" within much Arab Muslim culture. Though some intellectual innovations are regarded as contrary to Islamic law, others are accepted so long as they do not explicitly contradict those principles. Hence, for social work or social research methodologies the question at issue becomes one of where they stand in relation to the religious sciences. Thus there has been an active debate within those parts of the Arab world where sociology is a distinct discipline as to whether there should be an "Islamic sociology," in which the concepts of sociology are "Islamicized," and a parallel debate about whether social workers ought to be accountable to religious leaders.

Learning is traditionally carried forward through memorizing the Koran, based on mnemonic systems. Drawing on research carried out in parts of North Africa, Eickelman argued that "throughout the Islamic world" both religious and secular knowledge "are thought to be transmitted through a quasi-genealogical chain of authority which descends from master or teacher (*shaykh*) to student (*talib*) to ensure that the knowledge of earlier generations is passed on intact" (1978, 492). Insofar as learning is influenced by such traditional patterns, "understanding," unlike within Western traditions, is measured not by any ability to "explain" text but by the ability to use particular Koranic verses in new and appropriate contexts.

An almost automatic Western inference from this account may be that such learning systems will lead inevitably to a deadening of most senses of inquiry. Given the wider features of Arab Muslim learning, which include some informal peer learning, this is a risky assumption. However, the model of a man (*sic*) of learning also often is allied to accepted popular notions of social inequality as a natural fact of the social order and a restricted sense of social responsibility to criticize or change society.

Scholars are not in agreement on the extent to which the critical elements of this account should be attributed to either formal or lay versions of specifically Islamic culture. Indeed, it has also been argued that it is the *political* rather than *religious* environments of most if not all Arab regimes that are not conducive to either an open, interactive educational process or to a research-generated social critique of political systems (Massialas and Jarra 1983).[10]

Christianity and How to Think and Not Think

The general relation of Christianity to a history of social thought in general and the foundations and development of science in particular is an extensive topic (Harrison 2007, Shapin 1994), although it is one that has relevance to understanding lines of continuity and discontinuity with later social work. It was the Reformation that released science from the tendency to nature worship in Greek and Scholastic traditions and made science free.[11] Within the Catholic tradition, readers of a certain generation will be able to rehearse Biestek's principles of the casework relationship as readily as their times tables. I referred in chapter 1 to the philosopher Paul Helm's arguments regarding objectivity in science. In example 2.5 he illustrates the possibilities of a Christian view of science and faith.

EXAMPLE 2.5
Helm and a Christian View of Science and Faith

Theologians "are more like grammarians than like scientists or detectives" (32), teaching us how to think and talk and how not to think and talk about God. "They attempt to indicate the rules of intelligible speech about revealed realities" (32). The Christian epistemological starting point is that the Bible is God's revelation, but there are "blocks to knowledge and resistances. The human mind is a factory of idols as Calvin put it" (40). Hence, error may be attributable to a failure to investigate something properly, or it may be attributable more fundamentally to "antipathy to the data" (40).

"Our human knowledge is partial. . . . More important we are fallible in our knowledge. We make mistakes, even wilful mistakes" (44). "Only Scripture is infallible; interpreters of Scripture are fallible, partly due to their ignorance, but more significantly due to the workings of sinful biases" (44). Thus our knowledge is always open to revision. For those in the enlightened ages (David Hume's phrase) "we are living in an unprecedented time, the modern era," and "we judge the past, reinterpreting it in terms of the intellectual mores of modernity" (53). Helm suggests by contrast that "what we know about the past may exert decisive epistemic authority over the present" (53).

He takes the argument that there continue to be "rays of natural light" and understanding of first principles such that the testimony of the senses should be given proper regard. But we also need supernatural epistemic powers through the illuminating of the Holy Spirit. This does not give us new cognitive powers but enhances those we have. "To know of the true God the creator, special revelation and divine grace are every bit as much needed as they are in order to know God the redeemer in Christ" (59). Drawing on Augustine on *Nature and Grace*, "human nature is fallen but not thereby obliterated" (60). "If this were not the case, there could be no science, nor art, nor certainty in the nature of things" (61).

How does "science" then work for someone of faith? In the first instance, "our senses are to be used in a quite natural way" (63). There is no place for a scientific Gnosticism of discontinuous knowledge. "Grace builds upon nature, it does not ignore it or supplant it" (65). Hence the constant emphasis in the New Testament on the importance of the senses.

Source: Helm (2014).

Power, Politics, and Social Work Science

When we question whether knowledge reflects the external order of the world we ask, for example, how far knowledge is the result of the struggle for power. Foucault made this central throughout his life, even though his position shifted. In his early book, *The Birth of the Clinic* (1963, English trans. 1973), he traced the origins of clinical medicine in France from 1769 to 1825. Traditional histories, he saw, depict a period of superstition and blind reliance on ancient authority which were to be replaced by empirical science. Foucault argued that it was not new and more sophisticated insight but a change in how "illness" and the "doctor" were defined: "the new doctor started looking in a different way at a differently constructed object of scientific knowledge, namely illness" (O'Farrell 2005, 37). His focus was not only on how an object (for example, madness) is constructed but how knowers (doctors) are constructed. The potential relevance to social work seems clear. Unlike the prescriptions of most constructionist positions, we should consider not only how "evidence" is constructed but also how "evidencers" (social workers) are constructed within the social, political, and economic world.

Science attempts to make illness and those who are ill the object of orderly categories of knowledge; for example, as Foucault argued in *Madness and Civilization*, the mad become the mentally ill. The key term he introduced was "gaze," that is, how illness is observed and read by those taught how to "read" it. We might suggest by way of extension that there is an evidential "gaze" within social work—the way social work practice information is observed and read.

Where might we stand on the question of the relation of knowledge and power? I take a moderate standpoint position. A helpful perspective on standpoint positions can be achieved by revisiting a classic paper on the sociology of knowledge by Robert Merton. Merton analyzes the claims of this nature made by those who are epistemological *insiders* or *outsiders* to the group. The insider doctrine claims in its strong form that particular groups have monopolistic access to particular kinds of knowledge. In this form, the doctrine leads to the position that each group has a monopoly of knowledge about itself. In everyday language, "you have to be one to understand one" because "the Outsider has a structurally imposed incapacity to comprehend alien groups, statuses, cultures and societies" (1972, 15).

Applying this to standpoint epistemology, social work writers have often failed to distinguish strong and more muted versions of standpoint positions. The latter form of the doctrine claims that insiders have *privileged* rather than *monopolistic* claims to knowledge. This is a position that avoids the erroneous assumption of some radical advocacy researchers: that social position wholly determines what understanding is possible. As Foucault was to remark, late in his life, "there is always something in human experience that escapes the exercise of power," and "there is always a little thought occurring even in the most stupid institutions" (in O'Farrell 2005, 70, 71).[12] We pick this theme up in greater detail in chapter 8.

Science traditionally was seen as nonpolitical. In response to this, a critique developed of scientism, holding the general counterview that science is essentially political. The problem with this position is that when we say everything is political the argument becomes too broad and in effect means we no longer can make important distinctions. Often also such assertions leave the meaning of "politics" and "political" unspecified. One form of this is when some feminists say that "the personal is political" while not necessarily equating the personal with the political. The critique of scientism does have the value of recognizing the failure of a position that saw science as value free and also of a Marxist position that reduced science to economic factors. Hence "the claim that science is essentially political has helped open up science to both scholarly research and public critique and engagement" (Brown 2015, 10). Notions such as "experiment" or "survey" have long combined scientific and political meanings, seen, for example, in Campbell's "experimenting society" (Campbell 1969) and in the social engagement of the social survey movement in the United States almost a century ago.

But it is less than helpful. For example, it "tends to evoke polemical counter-claims." Also, "if something is always *already* political, it cannot *become* political." Then again, "in this broad sense, science is not only always political but also social, material, cultural, and linguistic." Finally, "conceiving science as essentially political neglects the importance of non-political relations for both science and democracy" (Brown 2015, 11). Something too rarely acknowledged in social work is that "calling something 'political' is itself a political act" (5). Politics is not only about power. Brown accepts a definition of "politics as purposeful activities that aim for collectively binding decisions in a context of power and conflict" that "includes four elements: power, conflict, purposeful activity, and the aim of collectively binding decisions" (19).

Doing Science and Doing Good Social Work Science

Doing science involves collaboration, contests, and controversy. We spend much of chapter 5 exploring aspects of these in social work science. The best of science takes in a level of mutual understanding rare in the ordinary life of research and teaching. Kuhn alludes to a colleague as "the only person with whom I have ever been able to explore my ideas in incomplete sentences" (1970, xi). Once again the tendentious contrasts too often made by some social work writers that stereotype the natural sciences look almost untenable when we hear the voices of scientists. We may hear at length the voice of Peter Mitchell, the biochemist and Nobel laureate, and find little that mystifies those doing social work science. Interviewed for the BBC by Lewis Wolpert (Wolpert and Richards 1997), and asked about having chosen chemistry, he responds, "I don't think I've ever really *chosen* anything much in my life . . . You recognize that there are *possibilities* there for your psyche, for your soul, for your being to fulfil yourself" (84).

Asked why he believes science can be done as well in a house or in the countryside as in an urban university, he reflects: "the ongoing process of science is really the gentle art of investigation into the nature of the world and ourselves. It *is* a gentle art" (86). He talked about how he got to his ideas. The problem he is basically interested in is the relationship between what goes on in a single cell and what goes on in the organism as a whole—the chemistry and the physiology. He wanted to find ways of understanding this as one thing, not two things:

> When I was first at university I became very interested in the Greek philosopher Heraclitus, and I began to think that there were two kinds of things in the world. One kind, which I called for myself "statids," were like teacups, which don't evolve in any way except that they gradually get broken, they get less recognizable. And the other sort were things like rivers, which Heraclitus talked about, which flow. The identity of the river is different, its environment is constantly going through it. So I called these things "fluctids," "flow things." Flames are like that and people are like that.
>
> (86)

The interviewer interjects, "But Peter, are you really telling me that your highly technical theory of how you get energy made in the cell was really based on these very general, and somewhat romantic images of the world?"

> Yes, completely so. And I especially agree with the word "romantic." I think this is something which more scientists ought to explain—that we don't do science because we are scientists, because of science—we do it because we are human beings. It is a most wonderful, romantic, cultural activity, just as much as being a sculptor, in fact, more so.
>
> (87)

He concludes with a reference to Jennifer Moyle, his collaborator, who since retiring had had a full artistic life. "Maybe that's been something about both Jennifer and me. I've often thought we're not really scientists at all, but we're just people, and we happen to have spent a lot of time doing science" (90).

Choice, art, sculpture, the interlinking of chemistry, physiology, and philosophy, the romantic, (un)certainty, and modeling reality all have parallels and echoes in science talk across the social sciences. By way of partial connection, here is a much briefer but possibly not altogether different social work voice:

> Being a professional social work practitioner or researcher is not a purely intellectual endeavor. It is an "art." I say "art" because there is between artists and their material a unique and special connection, as there is between scientists and their subjects. "Science" and "art" are not dichotomous. They are complementary. Both move away from the sensory to the realms of the theoretical and abstract in the pursuit of "truth." Both are enterprises of discovery, begin with observation, rely on similar forms of metaphor and analogy, go beyond nature, endeavor to rise above the literal, translate "data" into higher orders of conceptualization, plumb into deeper levels of significance, locate patterns and themes, decipher underlying meaning, and seek general truths.
>
> (Fox 2012)

Doing science poses further questions. What kinds of relationships exist between people who are doing science? How is doing science influenced by and interactive with other parts of our lives? Do scientists have epiphanies or, as Sinclair (2012) suggests, "slow revelations"? Are there research networks in social work science? Have there been—and are there—schools within social work in the sense that U.S. colleagues may have spoken of a Functionalist School or sociologists of the Chicago School? Taking networks as an example, a review of the state of social work writing and thinking on research networks leads one tentatively to conclude:

- Most of the articles that discuss networks are about action research or practitioner research—not mainstream social work research.

- The articles that speak of networks in social work are mainly about intentionally formed networks—sometimes with a capital *N*.
- There is no extended discussion of the nature of social work research networks.

The extract in example 2.6 is from a case study of social work research networks by Shaw and Lunt (yet unpublished), which illustrates something of the dynamics that might be present in an informal network held together solely by mutual research interests and with very low-tech support. One of the men in the network is speaking.

EXAMPLE 2.6
Network Dynamics

Well there's, there's a dynamic within the network which is applied versus basic if you like, or another way of putting it would be theoretically oriented versus practically oriented, and that's an important dynamic in the network, it's unresolved, and that's good. I mean it ought to be unresolved, because there isn't . . . a way of saying that Alexis's interest in Marxist social policy for example, Marxist theory, is irrelevant to the question of whether or not a welfare-to-work possibly could work in the Netherlands. There should not be, and there never was I think in the network, a position on that, because it was clear that there were theoretical perspectives which people could offer which would inform the way in which practical application of research to policy would take place, and for me, keeping that unresolved, valuing people's theoretical contributions, without advancing theory over practical relevance was incredibly important.

There was a wonderful exchange in [Finland] where a guy called Lucas Johansson attended from Sweden, taken over from Katrin, and John gave a highly theoretical account of a particular development, and Lucas said at the end of this, "But what use is this?" and the question was almost breaking the rules, because the point was it was, it was an offer from John to expand an area of thinking, and it's for network members to consider how it might apply in their case, or be useful to them, there shouldn't have been a request to John to say to any individual network member, "And this is what it means for your country or your research project or your specific research interest." But I think for me that's, that was an interesting example of the dynamic within the network that holds some things together in tension all the time, in tension.

Doing Work

Doing social work science involves doing work. Wherever we decide to stand, it will involve work. Foucault remarked that "to work is to undertake to think something other than what you thought before" (in O'Farrell 2007, 45). The cultural theorist Stuart Hall said more strongly, "I want to suggest a different metaphor for theoretical work: the metaphor of struggle, of wrestling with the angels. The only theory worth having is that which you have to fight off, not that which you speak with profound fluency" (in Grossberg 1996, 265–266).

In general I bring a view about science that in some but not all ways is like that of Weber. Science, he insisted in the gendered language almost universal at the time he was speaking (1919), demands a "strange intoxication." "Without this passion . . . you have no calling for science and you should do something else. For nothing is worthy of a man unless he can pursue it with passionate devotion" (1948, 135). As Medawar later expressed it, "to be good at science one must *want* to be—and must feel a first stirring of that sense of disquiet at lack of comprehension that is one of a scientist's few distinguishing marks." That entails "power of application and [a] kind of fortitude" (1984, 9, 10). This reminds me of a remark attributed to Isaac Newton and the story of how, when asked how he had arrived at the theory of gravity, said—not by seeing an apple fall but—"By thinking about it all the time."[13]

Going back to Foucault's remark, he does not quite seem to say that work is when we successfully end up thinking something other but when we endeavor to do so. (In context he was actually criticizing himself for one of his books, for having rested on his laurels). While writing this book I heard what I am willing to risk saying was a remarkable paper by Ed Mullen. Mullen has been a major figure in U.S. intervention research, thinking and writing mainly out of the mainstream U.S. commitment to the value and feasibility of what we may call scientific practice. He also has been part of a largely European network for about fifteen years. His paper had the working title "Reconsidering the 'Idea' of Evidence in Evidence-Based Policy and Practice."[14] I quoted this remark of Foucault to him and asked, "if, as a result of doing your paper, you thought something different." I give his reply here with his consent and because it nicely illustrates what I want to emphasize about doing scientific work.

> I selected the topic BECAUSE this was a genuine question that I had and I wanted to seek a new answer by exploring NEW literature. I had a great deal of curiosity about this topic and I found the traditional, established views wanting. I don't think I have worked so hard or examined such a broad array of new literature ever before when preparing a paper. Where I came out was totally unanticipated. I followed the ideas I came across and tried to synthesize them into a relevant and constructive new pattern. As I said, this is my synthesis and I expect that someone else might explore other or additional ideas and synthesize all of it differently. This synthesis works for me—pragmatically speaking. Indeed, this is a topic I have been thinking about and reading about for a few years now. I think Foucault's comment is a good characterization of my "work" on this paper. Of course, I probably would not (or could not) have done this if I was not retired and emeritus. My new status afforded me the time and the intellectual freedom and the curiosity to think anew.

Doing *social work* science connects with ways of understanding theory and practice, discussed in chapter 1, and cultures within social work practice. Fargion, in a valuable application to social work of ideas from the sociology of knowledge, identifies two cultures among practitioners engaged in fieldwork, seen as "embedded in different world views" (2006, 257), and relates them to Mannheim's work on styles of thought. Drawing her argument from fieldwork with social work practitioners, Fargion's contribution cautions against both generalization about the way social workers think and practice and also against simplistic distinctions between, for example, modern and postmodern worldviews. Mannheim takes the positions of the Enlightenment and romanticism and labels them as "conservative" and "natural law." An important feature of this argument is that what defined and distinguished these styles of thought was "not so much, or exclusively, their content as the way of thinking" (257). Thus, as Mannheim says, "Conservatism did not want to think something different from its liberal opponents; it wanted to think it differently." While Enlightenment thinkers wanted to present things in a coherent and rational framework, "the conservative mentality is . . . intrinsically resistant to strong systematization" (258). Romantic thinking focuses on concrete, material reality—"the present and the spatially circumscribed . . . as they appear directly to the observer" (258). Romantic thinkers are "skeptical towards science, and they interpret generalizations and laws as contingent products of our culture."

By contrast, "natural law thinking privileges the abstract; it transcends the actual present and focuses on abstract possibilities. . . . Events are perceived as epiphenomena, as if they were . . . accidental manifestations of an essence seen as independent from the context and history. . . . From this ensues the unconditional trust placed in science and scientific methods" (258).

When associated with how practitioners vary in valuing theory or practice, romantics give priority to being over thinking, practice over theory. "In a sense, even within a conservative style, thinking is just a kind of practice which is complementary, and not superordinate, to other practices" (259–260). By contrast, in natural law Enlightenment systems, thinking and being, theory and practice, are "different sorts of 'things' " (260). Theory and thinking are typically given a superior place over being and practice. Solomon aptly concludes that social work needs "thoughtful active caring" (Solomon 2007, 105)—but it will not be easy.

Insofar as this book is consistent, my convictions about what doing good social work science involves are manifested through all its pages. The word "good" is central here. For different readers it may mean good in terms of how adequately it reflects what we believe to be genuine science, good in terms of the moral and ethical aspects of the research process, or good in terms of the results and end in view from the scientific work. The recognition in these opening chapters of fallibilism, uncertainty, the limits of science, and so on warn us not to expect cut-and-dried answers. As the sociologist Harry Collins remarked a few years back, "Even if science is exact, it's exact only in the long term."[15] The canonical view of science as providing a neat and tidy path toward greater knowledge is a myth. Despite the brashness and gendered assumptions about "men" in science, C. P. Snow (1956) concluded tellingly that the greatest enrichment the scientific culture can give those within the traditional culture is a moral one. "Among scientists, deep-natured men know that the individual human condition is tragic. . . . But what they will not admit is that, because the individual condition is tragic, therefore the social condition must be tragic too. . . . The impulse behind the scientists drives them to limit the area of tragedy."

It is here that we reach a core consequence for the later scenes explored in this book. Good social work science will have something to say about what we know, about how we understand, and for its social purposes and outcomes. These are sharpened in chapters 6 to 8 through exploring what it entails by accepting that good social work science yields "evidence," fosters "understanding,". and has a commitment to "social and individual justice."

Taking It Further

Reading

Fargion, S. 2006. "Thinking Professional Social Work: Expertise and Professional Ideologies in Social Workers' Accounts of Their Practice." *Journal of Social Work* 6(3): 255–273.

Fargion's research is a helpful social work way into some of the issues in this chapter. Her own interests are transparently present.

Wolpert, L., and A. Richards. 1997. *Passionate Minds: The Inner World of Scientists.* Oxford: Oxford University Press.

I would suggest choosing one of the thematic sections and read through the full group of interviews in that section. Noting how scientists talk about ways of thinking and knowing about—and doing—science will be a constant reward, and I will draw from this source throughout the book.

Task

What is the adequacy of the arguments set out in this chapter? They can be critically assessed by bringing together at least one other member of a recent research team that had involved the reader.

In what ways did questions of personal experience, commonsense knowledge, creativity, power, or faith surface—or become suppressed—in the time the team were working?

Which aspects figured most weightily, and why?

[3]

Historical Moments for Social Work and Science

The questions focused on in this book have long antecedents, ones both expected and unexpected. After a brief introduction, I will suggest how I see the various positions from which the history of social work and research have been approached. I will offer some comments on what I see as the limitations of all or some of these five positions.

I suggest potentially interesting fields of study—interesting not only in substantive terms but for how they exemplify approaches fruitful for both social work and science. These are:

1. The history of science methodology
2. How fields and disciplines secure identity
3. Social work within wider contexts
4. What we may gain by thinking historically about inventions and discoveries, innovations and controversies, in social work

> Critical historians, like critical social scientists, now walk a delicate line between the empiricism that comes with evidence- and data-driven research and the relativism that comes with reflexive research. History, like social science, deals in materially grounded claims to truth and, as a result, deals in partial rather than absolute truths. It delivers partial rather than complete knowledge. It is rightly wary of rigid grand narratives and overly determinist explanations of social change.
>
> —P. Cox, "The Future Uses of History" (2013)

> What I always tell my students is that fossils are never what you expected. It's not possible, no matter what we connive and plan and try to predict what the past was like, it's never what we thought it was going to be.
>
> —Elwyn Simons, paleontologist[1]

Chapter epigraphs are largely absent from this book, but I could not resist the quotations at the head of this chapter. When we talk and think historically about science in social work, we are inevitably influenced by our presuppositions regarding social work and its history more generally. "Everyone knows who is traditionally said to have invented paper"—so might someone say in China. But almost no one in the United States or the United Kingdom knows.[2] By extension, what we believe to be the most important moments in the history of social work will vary if we ask someone in Japan, Germany, Sweden, Brazil, Egypt, Denmark, Italy, Australia, the United Kingdom, or the United States. Similar consequences would follow if we were to ask social workers from different countries what they think we should most regret or even apologize for in our histories, or if we were to ask who are the greatest characters in social work history. Indeed, what counts as social work and when it started will tell us something of what sort of story people come to accept. A national story of social work may be one of survival against political odds, or against the dominance of the "big" nation next door, or as one's own nation exercising imperial power. In this sense all stories of the past are really stories of the present. It may illuminate and surprise if we check out the people who appear on the History of Social Work website developed by Jan Steyaert.[3]

The term "social work" in its sense used today arrived later than we may assume. It is possible that the first occasion when the phrase "social worker" was used in something approximating its later sense of a professional or occupational role is in an 1889 paper by Mary Richmond on settlement and friendly visiting.[4] Richmond first asks, "What are the forces existing in and around the poor home with which the social worker must learn to work in sympathy and for which she must hesitate to furnish any more artificial substitute?" Richmond complements this role with a list of other occupations, and she refers to a "modern city's . . . general scheme of social work." But even when the term was used in something like our later sense, the composition of the occupational mix rings strange to contemporary ears. Eileen Younghusband, when recalling her association with the London School of Economics in the mid- and late 1920s, remarked,

> I remember the whole concept of what constitutes a social worker and where a social worker should be employed was extremely vague, amorphous at that period compared with what it is now. For instance, in

> many quarters personnel managers, women housing managers, youth employment bureau secretaries, were regarded as being social workers. Those were some of the employments into which social science students went.[5]

Speaking of the United States, Andrew Abbott lists some who were at the 1884 meeting of the National Conference of Charities and Corrections—"everything conceivably related to social reform . . . is contained in this meeting somewhere" (1995, 546). But by 1912 "we find a major change. . . . Many aspects of welfare-related work present in the earlier lists are going or gone" while other new links appear, such as psychiatry. "This set of connections is the great problem of social work history. Why was it that certain task areas became part of social work and other parts did not?" (546).

Having problematized "social work," we can identify five explanatory structures within which the history of social work and science has or can be framed.

1. Fall from grace or rise from the pit
2. Cycles of history
3. Science and social work as historically relative
4. Science and power
5. The story as told by the winners

Approaching the History of Science in Social Work

1. *Fall from Grace vs. Rise from the Pit*

Expressed more formally, these two notions set the romantic and Enlightenment views of history against each other. Indeed, it is possible to view much of social work's shared intellectual history and to a significant degree welfare interventions as a battle between those who wish to give emotion, spontaneity, intuition, and the life of the imagination their due recognition against those who rest their hopes in reason and the power of scientific progress. Given the centrality of these alternative positions to how science and social work are approached in our own time, I give more space to this than to subsequent frames of understanding.

By and large, social work's predecessors a hundred years and more ago held a far more optimistic view of science than do we, although the picture was not as monolithic as is often believed. In the United Kingdom that position has often been detected in the work of the Charity Organisation Society, but it was equally evident, though less frequently acknowledged, among many who backed the Settlement movement. Take Clement Attlee, later one of the most significant if least appreciated of British prime ministers. Deeply involved in the British Settlement movement, his book *The Social Worker* repays reading on this and other things (Attlee 1920). Example 3.1 extracts a series of observations he made about social work and science.

EXAMPLE 3.1
Clement Attlee on Science, the Settlements, and Social Work

On the emergence and success of social science: "The careful dissection and investigation of social phenomena is a comparatively modern achievement, but it has perhaps done more than any single factor to change the outlook of men and women on social problems."

"Research has been made into almost every phase of poverty, and many of its causes have been elucidated."

Science and social work: "Science has been rescued through the work of the practical social worker, the experimenter, and the investigator" such that science in this field has become "the hopeful science" and social work "the legacy of the prophets."

The "social investigator" is one form of social work "by cultivating habits of careful observation and analysis of the pieces of social machinery that come under his notice."

The scientific motive and social reform: "There are numbers of social workers who find in the work of research and investigation the best outlet for their desire for social service. . . . The scientific motive takes its place as one of the incentives that lead men to devote themselves to social service."

"Each group of social workers, each Settlement, has been a laboratory of social science in which new theories are tested."

Source: Attlee (1920, 14–18, 230).

This intertwining of science and reform seen in Attlee was almost a commonplace both in the United States and parts of Europe. Of Chicago it has been said: "All of social life was here and being investigated by sociologists" (Plummer 1997, 8). It was Robert Park (1929) who spoke of Chicago as "a social laboratory," and in a memorandum seeking support for the Graduate School of Social Service Administration Edith Abbott said, "The public social services call for the best that the university can give but the work is all of it new and demands social research and social experimentation" (Edith and Grace Abbott Papers, Box 19, Folder 1). Ada Sheffield anticipated that "case-work agencies . . . will gradually become what may be described as social laboratories" where "study of . . . cases would go on simultaneously with treatment" (1922, 38).

The science/reform nexus was connected to the influence of social evolution and British and American positivism. Burrow distinguishes the "speculative" from the "practical" students of sociology. Both were ethically inspired, and "no account of evolutionary social theory can do it justice which regards it merely as an attempt to apply scientific methods . . . to the study of society" (Burrow 1970, 93). Evolutionary social theory was an attempt to answer not merely the question "how does it work?" or, perhaps better still, "how does it happen?" but also "what shall we do?" (101). In case this is bracketed as a deceased and solely British preoccupation, evolutionary social theory and positivism in England "died . . . slowly and unspectacularly. . . . In philosophy the only Anglo-Saxon version of the recession from positivism with any pretensions to originality—pragmatism—was not English but American" (260).

We should not overstate the extent to which individuals, groups, and associations were aligned with one position or the other. Attlee, writing shortly after the Great War, expressed an important distinction as elegantly as anyone when he observed:

> Our attitude to social service will be different according to the conception that we have of society. If we regard it as at present constituted on the whole just and right, and approve of the present economic structure, social work will seem to us as it were, a work of supererogation, a praiseworthy attempt to ease the minor injustices inevitable in all systems of society. We shall see a set of disconnected problems not related to any one general question. On the other hand we may see as the root of the trouble an entirely wrong system, altogether a mistaken aim, a faulty

> standard of values, and we shall form in our minds more or less clearly a picture of some different system, a society organized on a new basis altogether, guided by other motives than those which operate at the present time, and we shall relate all our efforts to this point of view.
>
> (1920, 10)

Attlee's position is in some respects mirrored in Beatrice Webb's journal entries and later recollections. Her empirical science—too nuanced to elaborate adequately here—is reflected in her remark that "Every day actual observation of men and things takes the place of accumulation of facts from books and boudoir trains of thought" (1927, 185). Yet Webb came quite early to reject the religion of science. She expresses it tellingly in a diary entry for December 8 and 9, 1903, on Herbert Spencer's later years and how she had an increased "distaste for all varieties of utilitarian ethics, all attempts to apply the scientific method to the *Purpose* as distinguished from the *Processes* of existence." On Spencer she concludes: "His failure to attain to the higher levels of conduct and feeling has sealed my conviction in the bankruptcy of science when it attempts to realize the cause or the aim of human existence" (62). Indeed, already by 1887 she was writing in her diary that "the religion of science has its dark side. It is bleak and dreary in sorrow and in ill-health. And to those whose lives are continual suffering it has but one word to say—suicide" (115).[6]

All social reformers, Attlee opined, "belong to one or other of these schools of thought" (11). The Charity Organisation Society (COS) "takes its stand on the assumption that society as at present constituted is fundamentally just, that on the whole the distribution of wealth is not unfair, and that most people get what they deserve" (1920, 70).

One of the most unequivocal advocates of the role of science in social work was Arthur Todd at the University of Minnesota. His position combined a belief in the value of science, a direct connection between science and personal and social virtues, and an insistence that science is more an attitude of mind—a "scientific spirit"—than a set of methods or techniques. I do not believe Todd's position was held by everyone—someone who confidently knows "that no real scientist would forget the common decencies" (Todd 1919, 82) must have been some way from the real world even in 1919. But the very fact that this position could be held and extolled by a senior academic and published by a major academic publishing house is evidence enough for the proximity of such views to the social work and social science mainstream. Take, for example, the *Research Memorandum on Social Work in the Depression*, published in the 1930s, where the authors can say, almost as a given, that "one of the great values of scientific method is in the attitude towards life that it develops. It is an attitude of confidence that encourages effort as worth while in itself, as well as because effort is part of an indefinite future of attainment" (Chapin and Queen 1937/1972, 107).

"Science," Todd says, "promises . . . comfort and relief from loose thinking, looser talk, and the wasting of great reservoirs of valuable energy. It smacks of challenge and a trial of strength" (1919, 62). On the interweaving of science and virtue he insists science "does create the will to serve. . . . Science yields place to no other source of enthusiasm for social amelioration" (69). In a secular version of Paul's chapter on love in 1 Corinthians Todd says science is "broad, tolerant, earnest, imaginative, but poised and self-controlled. It is not impatient of contradiction and criticism given honestly and sincerely. It is fearless, truthful, teachable. It is able to withstand mob mind, sentimentality, sensationalism, and petty partisanship. . . . Finally, the scientific spirit means generosity, fellowship, and hearty cooperation untainted by jealousy" (73). He believed that in a sense "scientific research is a triumph over natural instinct, over that mean instinct that makes man secretive, that makes a man keep knowledge to himself, and use it slyly to his own advantage" (74).

On the scientific spirit he has much to say. He is fairly indifferent to the question whether social work is a profession, saying the real point is that "*the scientific spirit is necessary to social work whether it is a real profession or only a go-between craft*" (66). "Hence, in connecting science with social work, our aim is not so much immediate results as an attitude of mind" (71). Thus, while he readily concedes that "science is both an attitude and a technique," he cautions that "the technique of science is never fixed. Science always moves on" (85). He concludes that "social work . . . will become truly scientific only when every social worker sets as his ideal . . . knowing the truth as it is and adding to the sum of the truth for the creation of a world worth living in and working for. To work for the truth that shall make you free—that is the scientific spirit" (85).

> The scientific spirit does away with obtrusive personality; it pours a healthy astringent on one's ego. It broadens our sense of personality until we get the idea firmly fixed that we are merely representing the best

> thought of the community and are not exploiting our own vanity upon the poor and needy. This is a very subtle temptation and can only be met by rigorous scientific self-immolation.
>
> (81)

Standing opposite to this there can also be found doubts about science, though less often in writing. "This is partly because such a position is often viewed as essentially a moral one, which can be stated only with difficulty and argued not at all" (Timms 1968, 60).

Doubts of this kind, often heirs to the romantic tradition in thought and action (cf. Fargion 2006), can be observed when the therapeutically inclined Elizabeth Irvine bemoans that "science deals splendidly with all that can be weighed, measured and counted, but this involves excluding from the universe of discourse the intangible, the imponderable, all that cannot be reduced to statistics" (1969, 4); in more contemporary doubts expressed through feminist criticisms of masculinist methodology; or in some arguments for inquiry led by engaging critically in the political agenda. It is reminiscent of what Raynor calls the "myth of nostalgia" in the UK Probation Service, which sees the belief espoused "that everything was better in the old days when practitioner autonomy and 'established methods' provided all the guarantees of effectiveness that were needed" (2003, 340). At different positions between these poles can be found the more modest claims of Kirk and Reid (2002) and the early quotation from 1923, that "whether there be a science in all this or not, the problems are to be studied and solved in scientific ways—by openmindedness, by use of the teachings of experience, by efforts to see causes and results" (Lubove 1965, 142).

Social workers are, on some versions of this view, fallen angels (cf. Specht and Courtney 1994). It can be seen in the way the common response to almost every piece of government legislation on social work—in the United Kingdom at least—is met with cries of how the past values of social work have been betrayed. Carey (2014) has recently traced what he describes as the rise of cynicism among British social workers.

Burrow, in his important if little-known work on social evolution in nineteenth-century England, relates this to the discussion of the "savage" in the seventeenth and eighteenth centuries as oscillating between two views—a denial of his humanity and regarding him as an exemplar of human nature. The latter view—that of the "Noble Savage"—implies a favorable disposition—"a generous inclination towards unprejudiced cosmopolitanism" (Burrow 1970, 5). This is exactly parallel to ways we distinguish between nature and society or convention. Some forms of emancipatory social work are of this kind, with assumptions of presocial goodness and even rationality.

2. *History as Cyclic*

It is not easy to find this position expounded in the history of social work or research. It is often presented as a relatively pessimistic position, as in ideas about the decline and fall of empires, and in conversational exchanges where change seen by some as progress is dismissed by others as the return of past positions and arguments. However, such a theory does not necessarily deny the possibility of social progress.

Perhaps the most important social science advocate of a social-cycles view of history was Pitrim Sorokin (1889–1968), the Russian exile who ended up at Harvard. In his *Social and Cultural Dynamics* he classified societies according to their "cultural mentality," which can be ideational (reality is spiritual), sensate (reality is material), or idealistic (a synthesis of the two). He interpreted the contemporary West as a sensate civilization dedicated to technological progress and prophesied its fall into decadence and the emergence of a new ideational or idealistic era. There are perhaps traces of a simple version of this kind of view in remarks we saw more fully in chapter 1, made by Albion Small in 1924, then head of the Department of Sociology at Chicago. An archive minute records: "He called attention most interestingly to the fact that research moves in cycles and also that there are fads and fashions in research."

Explanations of this kind, at least in the West, have almost always been applied at the level of society or even civilization as a whole. I have not been able to trace a social work version, though it is possibly implicit in various arguments. For example, Davies's debated and significant argument about social work as often involving "maintenance" (e.g., Davies 1994) certainly seems to move away from a linear model of social work. Davies would place his position within the merits of systems theory and Mertonian functionalism for justifying the existence of social work. "I remain of the view that the underlying dynamic is concerned with the security of society, the social survival (or maintenance) of people deemed to be at risk and the provision of new opportunities for individuals to regain the initiative to further their own welfare."[7] Somewhat differently, there is a frequently heard argument about policy interventions that are initiated when relevant problem indicators such as crime waves are untypically high, which suggests that such interventions are likely to show apparently positive results because of the phenomenon of regression to the mean. This seems rather similar to an assumption that social indicators go through recurring peaks and troughs not unlike a cyclic model of change.

3. History as Relativizing

This can be seen in some forms of postmodernism; in the critical, post-structural interpretivism adopted by Denzin and Lincoln; and in the development of a strong paradigm position, with its idea of the incommensurability of knowledge between paradigms, as adopted by Guba and Lincoln. "In the social sciences there is only interpretation. Nothing speaks for itself" (Denzin 1994, 500). While this is not expressed as a view of the history of research, it denies the possibility of any linear view, of progress, or perhaps of decline. Hence, by truth, Denzin refers to "a commitment to post-Marxism and feminism with hope but no guarantees" (1997, 10).

Postmodernism, when covered in social work, most often refers to an approach to methods of inquiry, or to how one should think about, understand, and act in relation to particular subjects. Historical postmodernism is employed as a more general stance, which suggests that the social, political, and cultural organization of modernity has changed fundamentally and that therefore our world today is very different from what we have experienced in the past. This is typically deployed as an argument not only about what has happened but more significantly as a claim that postmodernism is something preferable to its predecessor.

4. History and Power

The ideas developed within critical theory were touched on earlier. Horkheimer, an early and formative speaker for critical theory from the 1930s, speaking of the relationship between thought and time, says that while there is an "essential relatedness of theory to time" (2002, 233) and that "history does not stand still" (234), this does not mean that critical theory has "one doctrinal substance today, another tomorrow" (234). "Every datum depends not on nature alone but also on the power man has over it" (244). Theory "never aims simply at an increase of knowledge as such. Its goal is man's emancipation from slavery" (246). Yet "to strive for all this is not to bring it to pass" (214). "The theory may be stamped with the approval of every logical criterion, but to the end of the age it will lack the seal of approval which victory brings" (241).

We should not place our hope in any one class to achieve this. "It is possible for the consciousness of every social stratum today to be limited and corrupted by ideology, however much, for its circumstances, it may be bent on truth. . . . The critical theory has no specific influence on its side, except concern for the abolition of social injustice" (242). Some recent critical theorists appear still more explicitly pessimistic—"there is no angel of history," as Bonefeld (2014) expresses it.[8]

Foucault's work draws on history as his tool par excellence as a means of analyzing and challenging existing orders of knowledge. There are several ideas that he developed at different periods in his work. In his *Birth of the Clinic* he dealt with the origins of clinical medicine in France from 1769 to 1825. Noting how traditional histories are framed within a period of superstition and blind reliance on ancient authority that is replaced by empirical science, Foucault argued to the contrary that the development of modern clinical medicine was not accomplished through new and more sophisticated insight but by changes in how "illness" and "doctor" were defined. "The new doctor started looking in a different way at a differently constructed object of scientific knowledge, namely illness" (O'Farrell 2005, 37).

His ideas of genealogy, archaeology, the history of the present, and power-knowledge have all offered fertile if sometimes hard-to-define ground. His idea of the history of the present has proved a fascinating idea for those wishing to apply his thought. He describes his work on a number of occasions as the history or the diagnosis of the present, as the analysis of "what today is." He notes that our own times and lives are not the beginning or end of some "historical" process but a period like, but at the same time unlike, any other. The question should simply be "how is today different from yesterday?" Caroline McGregor (formerly Skehill) has applied Foucault's notion to the history of child protection in Ireland (Skehill 2004) and has written one of the most helpful expositions of the "history of the present" for social work, illustrating how it is well suited to studies using a problematization approach (Skehill 2007, 450) and affords a "concern to use history as a means of critique of taken-for-granted 'truth' in the construction of practices and discourses" (451).

It is not easy to understand the difference between archaeology and genealogy (e.g., Foucault 2002), and this is "not made any easier by Foucault's own brief and less than enlightening comparisons" (O'Farrell 2005, 68). Most settle for the approximate idea that archaeology deals with discourses and genealogy with power. This is how I have tried to apply Foucault's thinking to the history of research methods in Chicago social work (Shaw 2015a). Archaeology, he concludes, tries to describe not the ideas, themes, thoughts, images, and so on but the discourse itself—a discourse as a practice that follows certain rules. To echo a remark made in chapter 1, discourse is not a "document" but a "monument"; it is not something to be interpreted. "It refuses to be 'allegorical'" (Foucault 2002, 15).

In his later work he generally tends to grant knowledge and power equal status, for example, through his notion of "power-knowledge," where each generates the other in endless cycles through their complex relations. "Power is not something that simply forbids and represses, but is something that produces particular kinds of knowledge" (O'Farrell 2005, 45). A variant of the focus on power is the idea that history is the story as told by the winners.

5. History as the Story Told by the Winners

To give just one example, the historian of sociology Platt complains that

> Accounts written from within sociology, as history of sociology, generally treat both other disciplines and groups outside the academy as parts of the background. They are seen as instrumental to the main aims of sociologists, or as introducing distortions into the . . . course of pure sociological development.
>
> (1996, 264)

The work of Jane Addams and the Hull-House Settlement is a case in point. The Hull-House Settlement was collecting systematic data before the sociology department was: "those methods and indeed topics, which were characteristic of the 'Chicago School' were equally characteristic of social workers and voluntary activists who were in the field somewhat sooner" (Platt 1996, 263).

Sociology's natural history focuses on the ideas and concerns of key thinkers; in contrast, social work history is seen as led by "the increasing realization of the need for professional expertise in the administration of aid and an expanding classification of the needy and dysfunctional" (Lengermann and Niebrugge 2007, 68). While social work's history is more than simply an expanding "classification" (Abbott 1995), the net result is the same—neither social work nor sociology are able to see the other as "made subjects." Once this view was established, "each would exist in the other's narrative typically as an absence, unseen and unreflected upon," having failed to construct "a sociological history of sociology" (Lengermann and Niebrugge 2007, 71).

Comments

What may we conclude provisionally from this sketch?

First, the dominance of English-language work means that too often the approaches are filtered through a U.S. lens. Flick, for example, contrasts the development of qualitative research methods in the United States and Germany, concluding, "in Germany we find increasing methodological consolidation complemented by a concentration on procedural questions in a growing research practice. In the United States, on the other hand, recent developments are characterized by a trend to question the apparent certainties provided by methods" (2006, 19). In Finland it was the discipline of social policy that (in the academia) took over social work and almost submerged social case work and social work education since the 1930 and 1940s. This lasted until the 1980s, when social work as a discipline of its own took off (cf. Lorenz 2004).[9]

Second, and less often noticed, most of these frameworks lead to a privileging of secularization rather than treating it as a matter for interest and understanding. While almost everyone would claim to reject Comtean images of progress, we do so rather selectively. We illustrated this from the accounts of George Herbert Mead and others in chapter 2.

Third, almost all the generalized frameworks we have touched on bring with them the danger of what Howard Becker called sentimentality, which we are guilty of "when we refuse, for whatever reason, to investigate some matter that should properly be regarded as problematic. We are sentimental, especially, when our reason is that we would prefer not to know what is going on, if to know would be to violate some sympathy whose existence we may not even be aware of" (Becker 1970, 132–133). The danger stems in part from the thrust toward a form of absolutism in each of these frameworks, an all-or-nothing choice. As Atkinson and colleagues comment in a comparable context, this "glosses over the historical persistence of tensions and differences." It "presents too orthodox a past" (2001, 3).

Linking this to the notion of a hierarchy of credibility, Becker (1963) distinguishes conventional and unconventional sentimentality. *Conventional* sentimentality is to prejudge that a belief must be true, better, and rational because it is mainstream and that a belief must be false, worse, and irrational if it is marginal or excluded. *Unconventional* sentimentality is the reverse of this. A belief is truer, better, or more rational because it is marginal or excluded, and a belief is false, worse, and irrational if it is part of the mainstream. There is an interesting echo of Becker's distinction between conventional and unconventional sentimentality when Merton says, speaking of nonconformists: "There is no merit in escaping the error of taking heterodoxy to be inevitably false or ugly or sinister only to be caught up in the opposite error of thinking heterodoxy to be inevitably true or beautiful or altogether excellent" (1971, 832).[10]

Who Is Working the Field?

In North America work has been done by Michael Reisch, Haluk Soydan, Karen Staller, Patricia Lengermann, and Adrienne Chambon. In Europe we should note the efforts of Walter Lorenz, Francisco Branco, Jan Steyaert,[11] Caroline McGregor (formerly Skehill), my own work, and two networks—a "Special Interest Group" on the History of Social Work and Research within the European Social Work Research Association and the Social Work History Network in the United Kingdom.[12] The older work of Noel Timms also merits resurrection (cf. Shaw 2014c). There have been special issues of journals on this general theme (*British Journal of Social Work* 2008, *Social Work and Society* 2011).

Fields for Study

How then might we identify and weigh the value of different arenas for historical study that will contribute not only in substantive terms but also in how they exemplify approaches that are fruitful for both social work and science? I will rough out four examples that to varying but limited extents already have been commenced, suggesting in doing so how they manifest approaches to doing social work science history that avoid the downsides often present in the approaches seen so far. These are:

1. The history of science methodology
2. How fields and disciplines secure identity
3. Social work within wider contexts
4. What we may gain by thinking historically about inventions and discoveries, innovations and controversies, in social work

In all of these a recurring presence is the reminder that it ill-befits us to regard social work as a self-contained field, practice, method, or set of purposes.

History of Research Methods

It is odd that almost no work has been undertaken on the history of research practices within social work. Speculatively, perhaps this is a consequence of the apparent general view that, while various arguments have been advanced that social work in terms of its values, aims, and intervention methods may be in some ways distinctive, no one seems to believe that social work engages in domain-specific research methods. Indeed, I have argued something akin to the second part of this position (Shaw 2007). A glance at sociology immediately highlights a difference, where work on this question has been done by a number of people. Some of this has interest for social work, for example, the numerous contributions by Jennifer Platt and by Martin Bulmer on the history of methods in the United States; the work of Ray Lee on, for example, the history of technology in the interview (Lee 2004); and Mary Jo Deegan's (1997a) work on the Hull-House papers. Turning to social work, there is some interesting work at a slightly broader level, for example, by Timms (1968) and

Kirk and Reid (2002), and some of my recent work touches on this (Shaw 2009, 2011b, 2014a).

Is this relative neglect reasonable? Accepting that social work research methods are not special to the field, it may seem quite right that attention has focused elsewhere. I want to dig down into this question through a rather unconventional approach, one that requires some uncommon reader-writer collaboration. Before reading on, please consider the exercise in example 3.2.

EXAMPLE 3.2
Social Work Research Methods and Practices

The following four extracts are taken from archives of different kinds and from a period ranging from 1923 to 1936. What are the key terms in each extract? What do you conclude about the beliefs about good research methods that may be held by each writer/speaker? ("Group research" refers to graduate students playing a central and collaborative role in the fieldwork.)

This task may be easier if undertaken with another colleague or student.

A. Two course outlines taught by the same person

"Social Statistics." This deals with "the study of the application of statistical methods to social problems, the collection and interpretation of statistics relating to pauperism, crime, insanity, feeble-mindedness, immigration, and unemployment."

"Methods of Social Investigation." This is "continuing the course in social statistics" and comprises "the methods of inquiry used in selected official reports and in the most important private investigations."

B. Report to a funding committee

Research on population and housing: "The material for Part I (Growth of the City) has been obtained through Library Research. The material in Part II has been obtained in various ways, especially by the study of documentary material and interviews with leaders in the various foreign colonies. The housing material in Part III has been collected through the method of group research."

C. Extract from a fieldwork account in a research monograph

"To get a picture of a cross-section of the West Side, it is easy to follow one of the north and south streets and go over the bridge over the river's 'south branch' straight ahead to the bridge over the 'north branch.'" Of the Old Lumber Yards district, "dilapidation was everywhere. The cellars, even the first floors, were damp because of the grading up of streets and alleys from three to seven feet above the level of the yards. The walls of the cellars and the floors of the first stories were often decayed and musty, with the water draining down about the foundations. . . . Floors were warped and uncertain, plumbing generally precarious . . . window panes broken or entirely gone, doors loose and broken, plaster caked and grimy, woodwork splintered and long unvarnished."

Then, "returning to the busy Halsted Street thoroughfare, one came to . . . " "Leaving the Lithuanian colony a journey was made around to the back of the great Stock Yards area . . . the area usually referred to as 'back of the yards,' where very congested and insanitary conditions are still to be found. . . . Looking down the narrow passageways, numerous frame shacks are to be seen on the rear of the lots."

D. Research and reform

"One question that immediately arose in the first house-to-house canvasses undertaken was the question whether to undertake to reform the various objectionable conditions which were found. . . . From the beginning, all our investigators were graduate students preparing for some form of social work. Unlike professional investigators or the students in a social science department in a university who are accustomed to 'observation with the idle curiosity of the scientist,' our investigators were accustomed to see 'what could be done about it.' Every effort was made to bring about an immediate improvement in any bad condition that seemed remediable."

Source: Shaw (2015a).

An immediately apparent observation about these extracts is that they seem to reflect a diversity of methodological approaches and preferences. The first extract focuses almost entirely on "statistics," but that recedes into the background in the second; the third presents a form of description familiar to what later would be called ethnography. Social intervention figures in the final extract. *Yet all four extracts are from the same individual.* The puzzle can helpfully be considered by considering the different audiences. In the first extract we have methodology as professional model (research practices as taught). The second brief extract was from a report to a university funding committee and is couched in the minimalist tones of selective accountability and claims making (reporting to funders). The third extract is of methodology as reported practice (taken from long research monographs), and the final one sets research practices in their relation to a vision for social service and reform.

Expressing this more generally, we may speak of the emergence of cultures of research methods as entailing discourses. Discourse in this context carries a fairly wide sense as a group of statements—groups of verbal performances that are not linked to one another grammatically (as sentences), or at a logical level (as formally coherent), or even psychologically (for example, as a conscious project), but at the statement level, as a way of speaking. They can be described and have sets of rules for how to speak. "One can define the general set of rules that govern the status of these statements, the way in which they are institutionalised, received, used, re-used, combined together, the mode according to which they become of objects of appropriation, instruments of desire or interest, elements for a strategy" (Foucault 2002, 129). However, a discourse "does not form a rhetorical or formal unity, endlessly repeatable, whose appearance or use in history might be indicated." It is made up of "a limited number of statements for which a group of conditions of existence can be defined." Discourse in this sense is not an ideal, timeless form. "It is, from beginning to end, historical—a fragment of history, a unity and discontinuity in history itself, posing the problem of its own limits, its divisions, its transformations, the specific modes of its temporality rather than its sudden irruption in the midst of the complexities of time" (131).

An exercise of this kind brings into focus and weight our collective "field of memory," enabling us to enunciate and describe aspects of an archive, the analysis of which

> involves a privileged region: at once close to us, and different from our present existence, it is the border of time that surrounds our presence, which overhangs it, and which indicates it in its otherness; it is that which, outside ourselves, delimits us. The description of the archive deploys its possibilities . . . on the basis of the very discourses that have just ceased to be ours; its threshold of existence is established by the discontinuity that separates us from what we can no longer say.[13]
>
> (Foucault 2002, 147)

Fields and Disciplines

We have hinted already, in speaking about history as the story told by the winners, that what come to be accepted as domains, fields, and disciplines represent the current collective majority view. Disciplines emerge as a consequence of negotiation and territorial claims—and this is as true of social work as of any other discipline (Shaw, Arksey, and Mullender 2006). They are neither intrinsic entities nor homogenous or self-contained fields of work. "What we today take for granted as the 'natural' division of social science into separate disciplines, for example sociology and social work, was a decades-long development" (Lengermann and Niebrugge 2007, 63) and, we might add, one that there is no reason to regard as final.

This process sometimes can be seen at work. To take a somewhat paradoxical example, Deegan's work has successfully and with due cause gotten under the skin of (male) sociologists, through her argument that Jane Addams and other women of her period and circles were as much sociologists as the men but were sidelined into social work. However, this seems to be just as much a "sociological" territorial history as that which she criticizes, in that it is a claim to something she would call the Women's Chicago School (of sociology).

Chambon (2012) has opened discussion regarding the disciplinary borders of social work in the United States and Canada. I have tried to understand the history of social work and sociology in the United Kingdom (Shaw 2014a), although through Alice Salomon and others I am sensitive to my relative ignorance of important work in, for example, Germany. My concern is not with setting the record straight but with the way we read history. Part of our problem stems from "reading history backwards" (Seed 1973, x)—a view of history that reads back distinctions (for example, "sociology" and "social work") that had less application in the nineteenth and early twentieth centuries than they do now. This stands alongside a tendency to see the divergence of sociology and social welfare "as simply an example of the division of labour concomitant with social progress" (Burrow 1970, 102).

> The history of the social sciences is still left largely a prey to the Whig interpretation of history. . . . The impression conveyed is that the history

> of the social sciences is to be written by isolating "anticipations" of modern concepts and methods from the irrelevant, because outmoded, theories of which they formed a part.
>
> (Burrow 1970, 19)

Burrow nicely remarks that "We may properly ask why we are not like our ancestors; there seems something odder about asking why our ancestors were not like us" (xxii). "A search for friends in the past" (19) is legitimate and interesting, but it easily slides into an exploitative use of history and leads to a tone that "is often harsh with the grinding of axes" (43)[14] through the rhetorical avowing and disavowing of central terms by interest groups. Fuller rightly laments that we have a misleading dual sense of history. On the one hand, "the history of the relatively distant past is *telescoped* so that knowledge-based movements from the past . . . are collapsed." This becomes "a strategy for legitimating historical amnesia. . . . Therefore, any awareness of anticipations of contemporary developments is bound to be lost. . . . On the other hand, for the history of the relatively recent past, events are *stereoscoped*: that is, a wedge is driven between two closely connected developments, making them appear to be on opposite sides of a fabricated divide" (Fuller 2009, 11).

This history of science is typically conveyed through textbooks. Kuhn, in his *Structure of Scientific Revolutions*, had little explicitly to say about teaching, but he seemed to slight it in his remarks about the role of textbooks in rendering scientific revolutions invisible. He reasons that the image of creative scientific activity is taken from textbooks. They inevitably disguise the role of revolutions such that

> the historical sense of the working scientist or the lay reader of textbook literature extends only to the outcome of the most recent revolutions in the field.
>
> Textbooks thus begin by truncating the scientist's sense of his history and then proceed to supply a substitute for what they have eliminated.
>
> (Kuhn 1970, 137)

Textbooks characteristically have a small section of history, either by way of introduction or by "scattered references to the great heroes of an earlier age" (138). This gives a sense of scientific history as cumulative and linear. Texts imply that "from the beginning of the scientific enterprise . . .

scientists have striven for the particular objectives that are embodied in today's paradigms" (140). It is thus extremely hard for us to see early social work as anything other than an earlier stage in the cumulative growth of the field and to see differences as "idiosyncrasy, error and confusion" (138). So "why dignify what science's best efforts have made it possible to discard?" Of the textbook he concludes: "More than any other single aspect of science, that pedagogic form has determined our image of the nature of science and of the role of discovery and invention in its advance" (143).

Research in Social, Economic, and Political Contexts

This is one area where rather more historical work has been done and where "social, economic, and political" does not exhaust possible contexts. Soydan (1999), for example, has written on the history of ideas in social work, and Reisch has written extensively on the history of social work from the framework of a social justice model for society (e.g., Reisch and Garvin 2015). We also saw in chapter 2 how social work and social work science are heirs to particular developments in faith-based positions. Social work technology also is a historical context. At any moment, for example,

> The way we collect, retrieve, store and manage data will always be shaped by context including the historic moment, and our relationship to it. . . . As we move between various constructions (such as book as "artefact" versus book as lending object, or Internet as study object versus functional tool) we change the nature of our own physical and social interactions with the object, with the environment, and with others.
>
> (Staller 2010, 287)

Andrew Abbott's work on how to understand social work in the light of its boundaries is interesting in this connection. Abbott is atypical: as a sociologist, early in his career he "began to find social work knowledge and psychological knowledge . . . interesting" and "just as interesting as psychiatric knowledge" (Abbott 1995, 548). The kind of historical work in a wider context is illustrated by research on the history of Israeli social welfare services by John Gal and Mimi Ajzenstadt (Gal and Ajzenstadt 2013, Ajzenstadt and Gal forthcoming).

Inventions, Discoveries, Innovations, and Controversies

These are potentially rich fields, but in social work they are almost untouched. To illustrate with an example elaborated in chapter 5, what do we make of the statement that William J. Reid "invented" *task-centered casework*? He was called thus, for example, in connection with his distinguished professorship in 1998, when it was recorded: "He is principally known as the inventor of the task-centered approach, widely recognized as pioneering a new method and philosophy of practice for social work, a field that had been steeped in long-term psychoanalytic practice prior to his research and writings."[15] How was *evidence-based practice* "invented"? Is it helpful to think of *child abuse* as something discovered or invented? Hacking, for example, develops and reflects on the meaning and significance of the fact that child abuse was not a constant. Thus, "the abusers' own sense of what they are doing, how they do it, and even what they do is just not the same now as it was thirty years ago" (Hacking 1991, 254). "No one had any glimmering, in 1960, of what was going to count as child abuse in 1990. . . . Since 1962 the class of acts falling under 'child abuse' has changed every few years, so that people who have not kept up to date are astonished to be told that the present primary connotation of child abuse is sexual abuse" (257, 259).

These areas are of interest theoretically (how we conceptualize an invention, discovery, innovation, or controversy), philosophically (for understanding how developments and dissent might be handled), and descriptively (through empirical work, in ways akin to the research that has been undertaken on science practice by sociologists of knowledge).

Take, for instance, Kuhn's work focusing on inventions and discoveries. He insists that a paradigm does not equal a pattern for replication but an object for articulation. To recall the synopsis in example 1.3, normal science *is* that articulation—a kind of (fascinating) "mopping up operation" (Kuhn 1970, 24). "Normal science" is not interested in inventing new theories and may indeed resist those who propose them. "Normal science does not aim at novelties of fact or theory and, when successful, finds none" (52). While he was hesitant to overdo the distinction, he did work with the idea that discoveries are of facts and inventions are of theories. But novelties constantly occur, and when "produced inadvertently by a game played under one set of rules, their assimilation requires the elaboration of another set" (52). He is not altogether happy with the word "discovery" because it assumes "discover" is like the word "see"—it happens at one moment and can be attributed to one person at one point in time. "Discovering a new sort of phenomenon is necessarily a complex event, one which involves recognizing both *that* something is and *what* it is"—it is "both observation and conceptualization, fact and assimilation to theory" (55) and so is a process over time.

Elaborating on the thought that discoveries and inventions take place over time, the problem is created of how an invention successfully sustains itself and continues to be "useful." Evidence-based practice can be seen in this light and also in terms of Foucault's historical limiting of discursive "statements." Example 3.3 is offered as a suggested way of understanding the widespread success of "evidence-based practice." Understanding evidence-based practice as an invention and as a discursive statement suggests that, to use Kuhn's term, "anomalies" may accrue over time.[16] Burrow's remark about nineteenth-century positivism can be applied to evidence-based practice. The positivist tradition was never monolithic nor invulnerable, "not only because of what was formidable in its opponents' case, but because of the inconvenient surprises which its own programme, the attempt to apply scientific methods to as wide a variety of social phenomena as possible, inevitably laid in store for it" (Burrow 1970, 1).

EXAMPLE 3.3

Evidence-Based Practice: Making an Invention Work

Evidence-based practice has successfully held sway for numerous reasons:

1. It offers a link to medicine through the key definition by Sackett and colleagues (1996) and therefore offers borrowed status.
2. It is readily transferable to and from other fields of applied study (education, health, social work) and so has become embedded within a bigger picture larger than any one field.
3. Whether it is true or not is not the whole of what counts as relevance. It can produce program packages that can be sold on to agencies. "What works?" seems to have commercial value (Collins and Pinch 1998, 88). "Commercial standards of proof can run in a very different direction to scientific standards."

4. It ties in to the work of the Cochrane and Campbell Collaborations, which gives it an infrastructure for development and diffusion.

5. It seems to offer a bridge between academic and "applied" work and seems to promise a way to avoid theory/practice tensions. It thus has the potential to interest policy and practice fields.

6. It is capable of varied interpretations and applications and so captures a number of issues. It takes on the flexibility—and imprecision—of what Popper termed a "bucket" category.

7. It links readily to related discourses, for example, the advocacy of the need for "scientific practice" in U.S. social work.

8. It is offered by its advocates as atheoretical and so requires less intellectual sophistication to grasp the basic motifs. As a consequence it is more easily taught in social work and professional programs.

9. It seems to travel well through *time* (a rhetorically plausible case can be made for a historical pedigree from as early as the Charity Organization movement, Mary Richmond, and the 1960s "What Works?" debates) and *place* (it is presented as culture free).

10. Its weaknesses of argument are clear, so it invites those "outside" to react to it by way of "correction" and thus stay within the frame of reference.

Kuhn's argument has implications in two ways. First, it shapes how we think of the research process. The very notion that any given field (for example, social work) is a domain "presupposes some conceptual scheme ordering the universe prior to the observation of relevant facts" (Hughes 1980, 63). This suggests that "the knower is an active constituent in the construction of knowledge." Theories are "inventions," not "discoveries." Appreciating this connects to the way that we study human beings, "who can also theorise about and have views on their lives as constituted in society" (63).

Second, it points to the value of exploring controversies—a topic we return to in chapter 5. "Controversies in and around science . . . are a strategic and fruitful site for science studies informed by a dual consciousness—epistemic and political" (Brante and Elzinga 1990, 33). The challenge arises from the difficulty of sustaining an unambiguous distinction between a fact and a theory—a discovery and an invention. Take statistics—a field where confidence in number may suggest we can escape theory. MacKenzie (1981) sets out the controversies that took place in British statistics between Yule, Galton, and Pearson. Yule and Pearson proposed alternative and competing measures of statistical association between two nominal variables—Yule's Q and Pearson's C. MacKenzie argues that we need to take account of the cognitive and social interests of the two parties. Pearson, he suggests, had a deep-rooted commitment to the utility of statistical prediction and developed nominal measures by analogy with interval level and regression theory. Yule is said to have a more pragmatic concern with the data itself and hence to have an interest in nominal data as phenomena in their own right. Pearson is also said to have an interest in measures of heredity relationships between generations; Yule had no particular interest in eugenics. Also Pearson was a typical member of the rising professional class and welcomed emphases that distinguished it from the working class; Yule came from a downwardly mobile traditional elite with generally conservative tastes.

McKenzie argues that if people are working in different traditions and with different goals, the "discovery" will take different forms. In response to the counterargument that this makes statistics arbitrary, he says that to see it as invention is not to deny "connections between statistics and the real world" (1981, 216). A contemporary statistician may well say "our statistics is different; we do not hold the eugenics position of Galton or Pearson." He responds, "To say this is false in one sense, true in another." False because it appears to claim that we have achieved, or been able to evaluate, a knowledge not structured by social context. It is true insofar as our knowledge has emerged at a historically different point. " 'Our' statistics is different from 'theirs' in that it has evolved from it; but, like 'theirs,' it is a social and historical product, and can and should be analysed as such" (226). While—as we will see subsequently in chapter 8—McKenzie's argument has been contested at some points, the part played by interests in fields of science that may seem theoretically neutral is inescapable, and historical research provides an ideal vehicle for that exploration.

Concluding Thoughts

I do not want to be caught in the romanticism I have disavowed earlier. There is, to repeat Werner Bonefeld's remark, "no angel of history." Yet this should not obscure the potential and interest of historical work for social work science. To note in closing, it has stimulated and been stimulated by developments in methodology. The huge program of library digitization has opened up a vast swath of archival materials. My own work on Chicago social work and sociology might not have gotten far off the ground without the serendipitous discoveries lying in the JSTOR digitized journals of the 1920s (Shaw 2009). The current revival of interest in archival and documentary methods (e.g., Shaw and Holland 2014, chap. 9) reveals various doors that we may be able to open.

Science work of this particular kind is deepened by Foucault's theorizing of the archive. We should not assume too concrete a notion of an archive. "It is not possible for us to describe our own archive, since it is from within these rules that we speak." We are engaged in a "never completed, never wholly achieved uncovering of the archive" (Foucault 2002, 146, 148). To end this chapter with a different point, work of this kind takes us to a less decided time, when social work was "social work" and sociology was "sociology," and so on. Hence there is the opportunity for conversations across the boundaries of history—a mutual participation in a history of the present for the social work domain.

Taking It Further

In 1980–1981 the late Alan Cohen conducted twenty-six interviews with some of the pioneers and formative players in the development of social work teaching and practice. They are now edited as an e-book. Although these are mainly British, the pleasure of reading and listening (the sound files are also available) will prompt innumerable thoughts about the history of social work and scholarship. They can be found at http://www2.warwick.ac.uk/services/library/mrc/explorefurther/speakingarchives/socialwork.

Kirk, S. and W. J. Reid. 2002. *Science and Social Work*. New York: Columbia University Press.

For a (slightly too) American historical perspective on science and social work Kirk and Reid (chapter 2) is a good read. The following sources also are suggested as providing ways of opening up less conventional ways of understanding science and social work.

Burrow, J. W. 1970. *Evolution and Society*. Cambridge: Cambridge University Press.

Lengermann, P., and G. Niebrugge. 2007. "Thrice Told: Narratives of Sociology's Relation to Social Work." In *Sociology in America: A History*, ed. C. Calhoun, 63–114. Chicago: University of Chicago Press.

Lorenz, W. 2007. "Practising History: Memory and Contemporary Professional Practice." *International Social Work* 50(5): 597–612.

I generally have avoided recommending my own work for further reading, but because I have written fairly extensively on themes covered in this chapter, and in ways that question much social work history, I make an exception here and suggest:

Shaw, I. 2014. "Sociology and Social Work: In Praise of Limestone?" In *The Palgrave Handbook of Sociology in Britain*, ed. J. Holmwood and J. Scott, 123–154. London: Palgrave.

Shaw, I. 2015a. "The Archaeology of Research Practices: A Social Work Case." *Qualitative Inquiry* 21(1): 36–49.

Shaw, I. 2015b. "Sociological Social Workers: A History of the Present?" *Nordic Journal of Social Work Research* 5, sup 1: 7–24.

Shaw, I. Forthcoming. "Case Work: Re-forming the Relationship Between Sociology and Social Work." *Qualitative Research*. DOI: 10.1177/1468794114567497.

[4]

Technology and Social Work

Here I start from the previous chapter's attention to social work science and history and move from a wide-angle view to look at specific issues and challenges for social work. A brief excursion regarding the term "technology" will lead us into arguments for and more often against the nature and consequences of what is typically seen as a growing role for technology. The focus then moves to the relationship of technology to society as a whole, to science in general, and to professional knowledge and practice.

The later portions of the chapter explore a series of questions central to technology in social work. How straightforward is it to interpret evidence from technology? How might a distinction between experiments and demonstrations inform social work thinking and practice? Has the expansion of technology enhanced the possibility of collaboration in social work science? What future issues face social work technology?

Ada Sheffield, as with many of her contemporaries, had much to say about recording social work, including a sensitivity to shifts in technology. She devoted a chapter of her book on the social case history to writing the "narrative"—the client story—which "most reflects the case worker's skill" (1920, 75).

In a section on "The Typewriter and Narrative Standards" she talks of how the advent of the typewriter led to the narrative history being longer and assuming greater importance. "Indeed, it is a question whether we should today be thinking about record keeping as an expression of social case work, were we still held in bondage to pen and ink" (75). She recognizes that "labor saving devices" like the typewriter never simply enable easier accomplishment of a previous task—they always transcend earlier levels of experience and valuation. "The typewriter is bringing about a change even in the subject matter of our social case histories" (75).

She sees this as positive and compares a handwritten record from 1899 with one from a then contemporary record. The "pen and ink record is, as appears from the excerpt, terse and objective. It shows the worker to have been 'on the job' but concerned solely with relief giving " (79). While concerns with personality, then current, are accepted as a reflection of the development of social science trends, by comparison with handwriting "the typewriter is releasing time and energy in ways that count for the enrichment of our thinking upon our clients' problems" (80).

However, despite her general sense of social progress and commitment to "science," she does not take a naïve position. For example, she sees one consequence of the typewriter as being that "dictation to a stenographer lapses into the prolix and redundant style of ordinary talk" (81). Issues of fact selection would not have arisen in same way in the days of pen and ink. Changes in technology had introduced ephemeral matter. She delightfully calls some of this "behold-me-busy details," where narrative is "constantly interrupted by what is virtually an accounting to the supervisor . . . for time spent" (82). "In each instance the important concerns of the client are dropped for a space while the visitor makes it clear that she is on the job and earning her pay" (84).

Chicago School sociologists were beginning to advocate recording interviews in the actual worlds of the speaker (Lee 2004, Shaw 2009) and were using the expression "life history" almost interchangeably with "interview." This entailed the use of amanuenses and in the 1930s to the employment of stenographers. For a brief period in the 1940s a phonograph was used, but in the 1950s the advent of tape recording took over, such that by the 1970s this had become standard. The later advent of visual methods was largely made possible by subsequent developments in film and phone technology, although a glance at journal articles and books of the 1920s would show that images were far more commonplace than they were half a century later.

We tend to talk about "new" technology. While there are new *forms* of technology, the phrase "new technology" may be misleading. What we think of as "new" is not essentially so but only so at this moment in our time. As Karen Staller says when talking about this topic, "the future quickly becomes the past and that which is cutting edge today will be—someday soon—antiquated" (2010, 287). This is partly because "the way we collect, retrieve, store and manage data will always be shaped by context including the historic [*sic*] moment, and our relationship to it" (287).

The Word

Polkinghorne says that the Greek *techne* "is the knowledge and skills needed to protect oneself from the suffering that nature can inflict" (2004, 10) and includes counting, articulate speech, and medicine. It enables us to farm, build, and tame animals—"the collection of skills by which humans exercise control over nature" (11). Williams (1983) notes that by the early eighteenth century "technology" referred to a systematic study of the arts, "especially the Mechanical." It was only with the specialized sense of "science" that technology became fully specialized to the practical arts. "Technique" and "technology" often had similar uses, except that technology (from "-ology") also had the sense of systematic treatment.

There were early references to the seven liberal arts, and "artist" referred to almost any skilled person, thus in effect being almost identical with "artisan." Specialization of the term to painting, drawing, and sculpture (and eventually musical composition) was not fully established until the late nineteenth century, hence our distinction between artist ("fine arts") and artisan (skilled manual worker). There was a later idea of the "useful arts," which in time was taken over by "technology."

Though this is the root of our "technology," the term has taken on other meanings, such that the original notion of the study of skills has faded. We also use it in sense of "the particular knowledge and skills that are based on modern science" (Polkinghorne 2004, 11), thus bringing technology and science relatively close. It sometimes carries the active sense of transforming nature. The diversity of the term widens further when we often use the word to refer to the things themselves, that is, the devices produced by such skills and knowledge.

Polkinghorne is more concerned to make a point when he says that *techne* is a way of seeing nature as something to be used and controlled: "technology is a way in which things appear; it is a revelation" (40). He says, quoting Heidegger, that while it is "correct" to say that technology is a means and a human activity, it is not "true" because it misses the essential. "The essence of something is not the object itself, but the way it pursues its course; the way it remains the way it is through time" (39–40).

Doubts and Fears

It will be clear already that, with the refreshing exception of Raymond Williams, it is difficult to find statements about technology that are intended to be free from valuations one way or the other. In the frequent use of the phrase "new technology" we can see how "technology," like "science," functions as a positive rhetorical value—an ideograph (Rip 2002, 106). Martinez-Brawley and Zorita are among the few who seem relatively sanguine about the advent of technology in social work, echoing in their writing traces of continuing Enlightenment views. Social work education and practice have moved, so they suggest, from "preparation for a cause or an artistic undertaking" based primarily on tacit knowledge to "preparation for a technological undertaking or market endeavour which requires codified knowledge" (2007, 534). This is still some way from the brio of a famous UK prime ministerial speech half a century ago from the days of Harold Wilson's Labour government regarding the "white heat of technology"[1] and from Hitchings's remarks about the Great Exhibition of 1851 in Victorian Britain, seen by apologists such as Charles Babbage as "a chance to give the intricacies of technology an aura of religious magnificence" (Hitchings 2008, 277).

More neutrally, Rip refers to how technological developments may fragment disciplinary boundaries. "Disciplinary boundaries become less important in genomics and nanotechnology, but also in earth and environmental sciences" (2002, 99). He also observes, "one striking phenomenon" is "how information and communication technologies can address issues of indigenous knowledge, for example in making layered access to knowledge possible" (99).

But one does not need to turn to the more severe critics to observe cautiousness. Kirk and Reid, for example, note that "there is little recognition that scientific technology has limits or that what researchers have labored to produce may not be particularly usable" (2002, 190). Again, C. Wright Mills voiced criticism of the idea of "cultural lag": "The model in which institutions lag behind technology and science involves a positive evaluation of natural science and of orderly progressive change. Loosely, it derives from a liberal continuation of the enlightenment" (1943, 177). Taylor (1999) offers a mediating position through an interesting variant idea of modernity, globalization, and forms of technology, arguing that the manner in which various cultures adopt scientific practices differs, resulting in multiple or alternative modernities. This is in part because cultures do not start over when they adopt modern views; there is continuity with the past, and hence there is both convergence and divergence.

The more deeply held doubts cluster around three overlapping criticisms: that technology produces a calculative thinking, that it atomizes society and risks new powerful elites, and that it reflects social forms marked by surveillance.

Calculative Thinking

Polkinghorne, speaking of teachers, nurses, psychotherapists, and social workers, detects a trend to "substitute a technologically guided approach for determining their practice in place of their situationally informed judgments" (2004, 1). He sees this as part of a more general move to "a technified worldview," which "occurs when the technological worldview comes to permeate all the sectors of a culture" (25). Van Maanen says something similar:

> In professional fields such as pedagogy, psychology and nursing, the dominance of technological and calculative thought is so strong that it seems well-nigh impossible to offer acceptable alternatives to the technocratic ideologies and the inherently instrumental presuppositional structures of professional practice. The roots of this technologizing of

> professional knowledge have grown deeply into the metaphysical sensibilities of western cultures.
>
> (2007, 19)

He notes how "our technological understanding of being produces a calculative thinking that quantifies all qualitative relations, reducing entities to . . . programmable 'information' " (19).

Atomized Society

E. M. Forster's 1909 short story "The Machine Stops" is unsurpassed as a classic tale about the dark side of technology. I have long thought it prescient. The story's narrator remarks that in the Age of the Machine "Men seldom move their bodies; all unrest was concentrated in the soul" (Forster 1954). Hitchings laments "the solipsistic self-pities of our atomized society, and social fragmentation that results in linguistic separation" (2008, 328). Taking science and technology controversies as examples Brante and Elzinga (1990) ask if there is an inverse relationship between the power of experts and democracy and express concern over whether a technological elite is taking over.

Surveillance

Foucault, in his explorations of the relationship between knowledge and power, spoke of disciplinary power. This is for Foucault a "technology" aimed at "how to keep someone under surveillance, how to control his conduct . . . how to improve his performance . . . put him where he is most useful" (in O'Farrell 2005, 102). It involves organizing space in a particular way with enclosures within which smaller partitions are established (classrooms, wards, cells) and activities organized. He developed the idea of panopticism, where people can be observed but cannot see back—an inspecting gaze exercised through the examination of individualized cases. Social work assessment in residential assessment units can be explored in this way.

One of the most sustained social work expositions of this notion of panopticism has been developed by Nigel Parton (e.g., Parton 2008a, Parton and Kirk 2010). With Kirk he sets out how "those considered potentially at risk are the subject of increased state surveillance, intervention and control, even though many, if not most of them, are not and will never become 'cases' of the problem" (Parton and Kirk 2010, 29). They speak of the "de-emphasis of the relationship, the abandonment of explanation, and the growth of surveillance" (32) such that while the technical requirements of the job have increased, space for professional judgment has decreased, and "social work becomes increasingly involved in ever wide-ranging, complex and unstable systems of surveillance, particularly where such systems are used to enhance strategies for early intervention" (33–34). I quote at greater length:

> With the introduction of ICT there is an expectation that as information becomes more accessible, the agencies, professionals and their decisions should become more transparent and accountable. In the process, there is less discretion for the individual professional for identifying what information is relevant as the required information is predetermined by the structure of the database and the algorithm. The identities of clients as people with needs and problems in contexts are superseded by accounts constructed by the fields that constitute the database. . . . In the process, the embodied subject is in danger of disappearing and we are left with a variety of surface information which provides little basis for in-depth explanation or understanding.
>
> (33)

Later in the chapter we will return to this argument, questioning the strong position of technological determinists who hold that technology gives direction to other aspects of culture. For the moment, we should note that just beneath the surface of several of these anxieties the familiar philosophical criticism of "dualism" can be observed, alongside a tendency to nostalgia and sentimentalism for that lost time when "everything was better in the old days when practitioner autonomy and 'established methods' provided all the guarantees of effectiveness that were needed" (Raynor 2003, 340). Heidegger thought that older "handwork technology" worked with nature while modern technology takes things from nature. It "challenges nature. It commandeers energy from nature. . . . Modern technology . . . approaches nature aggressively, attacking it head-on, depleting its resources, draining its energy" (1977, 2, 6–7). "Man, investigating, observing, ensnares nature as an area of his own conceiving . . . an object of research" (9). "The essence of modern technology is its enframing of the natural world. . . . The threat of modern technology is not from its potentially lethal machines and apparatus; rather, its threat is to drive out every other possibility of openness to being. . . . Enframing means that way of revealing which holds sway in the essence of modern technology and which is itself nothing technological" (10). Yet Heidegger also believed we are not fated to remain within the confines of modern technology—we do so by expressing yes yet no to it. He did not advocate abandoning technology. He thought we can affirm the unavoidable use of technical devices and also deny them the right to dominate us.

Much of this worry about technology is in fact less about technology and more about society, the citizen, science, and professional work.

Technology and Society

David Runciman, writing in the *Guardian*, captures several recurring themes in thinking about technology and society, in ways that recall the fears and doubts outlined in the previous paragraphs.[2]

> The foundations of the information technology revolution were laid during the cold war. It has its roots in the massive US government research and development programmes of the 1950s and 1960s. During this period most of the money spent on scientific research in the US came out of the military budget. That spending was fuelled by cold war paranoia.

Furthermore, while the Internet began life as a military project, "so too did text messaging." He speaks of how "networks of people with shared interests, tastes, concerns, fetishes, prejudices and fears have sprung up in limitless varieties," and now "advances in computing have thrown up fresh ways to think about what it means to own something, what it means to share something, and what it means to have a private life at all." But "it would be a mistake to overstate what phones can do. They won't rescue anyone from civil war. . . . In the end only politics can rescue you from bad politics." He notes that the dawn of the information technology revolution of the 1990s raised hopes that these new communication technologies would spell doom for authoritarians because they would not be able to control them. "That is not what has happened. The Internet has not democratised the Chinese state. Instead the Chinese state has used it to bypass democracy."

For Max Weber scientific progress is part of the process of intellectualization. Does this mean we know more about life than "an American Indian or a Hottentot?" Hardly. For example, "the savage knows incomparably more about his tools" than do we about the technology that surrounds us. Weber viewed it as a process rather like secularization—"that one can, in principle, master all things by calculation. This means that the world is disenchanted" (Weber 1948, 138). Weber, in analyzing the effect of technology and industrialization on society, identified four types of human action:

- Instrumental rationality, whereby we seek to gain ends by calculating how people will behave
- Value rationality, through which actions are done because they are the right thing to do, that is, for ethical, aesthetic, or religious ends
- Affectual actions, which arise from, for example, anger or love
- Traditional actions, which are engrained as habits in the actor

Weber held that the first kind of action, instrumental rationality, was predominant in the Germany of his day. Value-rational actions had become devalued, as had traditional undergirding systems of belief, and so in that form of action there was no longer a code of ethics to direct people's actions. He believed that a process of rationalization had brought about the transformation of Europe into a modern society.

While "globalization" as a term has become commonplace only in the last two or three decades, the idea is not new. Consider the emergence in the nineteenth and early twentieth centuries of high-speed travel (rail and air) or communication (telegraph or telephone). Writing in 1839, an English journalist commented on the implications of rail travel by anxiously predicting that as distance was "annihilated, the surface of our country would, as it were, shrivel in size until it became not much bigger than one immense city." A few years later, Heinrich Heine, an émigré German-Jewish poet, captured this same experience when he noted: "space is killed by the railways. I feel as if the mountains and forests of all countries were advancing on Paris. Even now, I can smell the German linden trees; the North Sea's breakers are rolling against my door" (in Schivelbusch 1977). In Marx and Engels's account in *The Communist Manifesto*, the imperatives of capitalist production inevitably drove the bourgeoisie to "nestle everywhere, settle everywhere, and establish connections everywhere." But despite their ills as instruments of capitalist exploitation, new technologies that increased possibilities for human interaction across borders ultimately represented for Marx a progressive force in history.

Heidegger also accurately prophesied that new communication and information technologies would give birth to novel possibilities for dramatically extending the scope of *virtual reality*. Writing in the late 1960s on "What Is a Thing?" he said, "Distant sites of the most ancient cultures are shown on film as if they stood this very moment amidst today's street traffic. . . . The peak of this abolition of every possibility of remoteness is reached by television, which will soon pervade and dominate the whole machinery of communication." In his analysis, the compression of space increasingly meant that from the perspective of human experience "everything is equally far and equally near." The abolition of distance tended to generate a "uniform distanceless" in which fundamentally distinct objects became part of a bland homogeneous experiential mass. The loss of any meaningful distinction between nearness and distance contributed to a leveling down of human experience.

Gambrill draws on a different source to ground her ambiguity toward technology, from her reading of Jacques Ellul's work. Ellul was a philosopher who approached technology from a dialectical viewpoint, the dominant theme of which was modern technology's threat to human freedom and Christian faith. His constant concern was the emergence of a technological tyranny over humanity. Gambrill picks up his idea of a "technological society" (Ellul 1964), a theme that has close connections with later work referred to above on the idea of an "information society" (Parton 2008b). This, she says, "is the technological society, dominated by the mass media" (Gambrill 2010, 306), and she relates that, following Ellul, to the political form of society and her view that Ellul "argued that the effects of propaganda are always negative (especially in a democratic society) whether intentional or not" (313). Her response is that propaganda is an essential (though she seems to intend "inevitable") part of society, so we cannot escape its influence but must aim to understand it so we can "mute its effects" (314).

Technology and Science

"Distance lends enchantment. . . . Scientific and technological debates seem to be much more simple and straightforward when viewed from a distance" (Collins and Pinch 1998, 2). Yet ironically "the greater one's direct experience of a case, the less sure one is about what is right" (3). Even tales of technological heroism are tales of human endeavor rather than superhuman feats.

Technology is used and demonstrated in circumstances less controlled than mainstream science. But science cannot rescue technology from doubt: "when looked at closely the conditions seem as wild inside the lab as outside" (Collins and Pinch 1998, 3). "It would be foolish of course to suggest that technology and science are identical" (4). Technology is closer than science to the worlds of politics, business, and the professions. But they are differences of *degree*: "Since all human activity takes place within society, all science and technology has society at its centre" (5). Leroy Hood, a biotechnologist, likewise rejects a gulf between technology and science and sees what he does "as this beautiful fusion between biotechnology on the one hand and leading edge biology on the other" (Wolpert and Richards 1997, 38).

Technology and science constantly interact with each other. Scientific discoveries can form the base for new professional practice, but the opposite also holds; that is, in retrospect, technological innovations "from the floor" are analyzed and theorized by science (Brante 2010, 859). Indeed, "every scientific issue . . . is made more pregnant and vivid by seeing how it emerges in the setting of electronic communication in all its varieties" (Benthem 2002, 72). "Technological and broader scientific issues work together in generating many new fundamental issues for a natural conglomerate of sciences of information and cognition" (74).

More specific and helpful for social work science is a deeply significant argument by Stokes (1997), and it is one that I deal with partly here and more explicitly in the final chapter. His general focus is on the nature of the link between the drive for fundamental understanding and the drive toward applied use (or, as he says, technological innovation), and he pleads for a more realistic understanding of the relationship between science and technology.

It has been fairly readily acknowledged that the linear model oversimplifies the pathway from science to technology. Stokes challenges two important dubious assumptions about the commonly held view of a linear relationship between science and technology, between basic and applied science. First, that technology innovation is rooted in science. He counters this by saying that for most of human history—and still today—the practical arts have been perfected by "'improvers' of technology" (1997, 19) who knew no science. But second, he says, "the deepest flaw in the dynamic form . . . is the premise that such flows as there may be between science and technology are uniformly one way, *from* scientific discovery *to* technological innovation; that is, that science is *exogenous* to technology" (19). He suggests that this has always been false in the annals and experience of science, and it has on occasion been exactly the reverse, in cases where scientists have learned from trying to understand technology. But more subtly, developments in technology become a far more important source of the phenomena that science undertook to explain. Science has structures and processes that exist only in the technology, such that science becomes technology derived.[3]

There was a strand in Europe that saw science as about controlling the world and not only understanding it. This was apparent through the influence of the Christian tradition, with its very different view of manual labor, and of science as a Christian responsibility in response to the Fall. To Francis Bacon science and technology were blurred: techniques *were* knowledge rather than the fruits of knowledge, and this view resulted in the foundation of the Royal Society with its utilitarian aims. We saw in the opening chapter how Hooykaas approvingly quotes Thomas Spratt, the eighteenth-century author of *The History of the Royal Society of London*: "philosophy will then attain to perfection, when either the Mechanic labourers shall have philosophical heads, or the Philosophers shall have mechanical hands" (Hooykaas 1972, 96). Stokes nicely summarizes the death of the linear model: "It has been dealt mortal wounds by the spreading realization of how multiple and complex and unequally paced are the pathways from scientific to technological advance; of how often technology is the inspiration of science rather than the other way round; and of how many improvements in technology do not wait upon science at all" (1997, 84).

He wants to see technology and science "as interactive but semi-autonomous" (87). The significance of his distinctions between pure basic research, use-inspired basic research, and purely applied research and development is explored in chapter 9.

Technology and Professional Knowledge and Practice

The diverse forms of interaction between science and technology that we have observed have consequences for professional practice. For example,

Kuhn acknowledges that a gap in his *Structure of Scientific Revolutions* was the lack of attention to technology, but there are moments where his argument illuminates such thinking and practice. When discussing the emergence of anomalies within paradigms he notes that the usual response to anomalies is to wait and not go for a new paradigm. In 2011, for instance, there was a suggestion that neutrinos at the CERN Large Hadron Collider had traveled faster than light, which would have had profound implications for the whole of physics. This led to a subsequent checking of instruments, followed by the conclusion it was a measurement error and, significantly, the resignation of the scientist who had reported the phenomenon.[4] But *resistant* anomalies shift the focus of a discipline and even change how scientists see the form of the field. To underscore a point made earlier in this book, early-paradigm research will follow the rules quite closely, but increasingly the paradigm gets modified even to the stage where it is not agreed what the paradigm is. It is possible to see parallels to this in the field of technology and social work through what we may call "policy engineering." Thus in a UK study of the development of electronic records for children's services, one pilot site displayed an early commitment to following the recording format ("Exemplars") as given, but the site managers gradually modified the format to fit local circumstances and respond to their sense of anomalies in how the system worked, although doing so within a recognizable version of the paradigmatic template (Shaw et al. 2009).

An analogous point can be made from Giddens's discussion of aspects of the relationship between the social and the natural sciences. He insists that he does not set up any prohibition of interaction between science and lay culture, but while common sense may and rightly does shape the way hypotheses and problems emerge in the natural sciences, "the relation between the natural scientist and his or her field of investigation . . . is neither constituted nor mediated by mutual knowledge, in the way I have defined the term" (1993, 14). He says this is not undermined even by extreme constructivism, for "no-one suggests that it is the natural world which constructs accounts of itself" (15).

He makes an interesting inference that one consequence "is that original ideas and findings in the social sciences tend to 'disappear' to the degree to which they are incorporated within the familiar components of practical activities. This is one of the main reasons why social science does not have parallel 'technological' applications to natural science." Viewed in this way, "the impact of social science—understood in the widest possible way, as systematic and informed reflection upon the conditions of social activity—is of core significance to modern institutions, which are unthinkable without it" (15).

This seems to lead to a further inference that professions will vary considerably in their relationships to technology. Brante works this point out from his view that professions are "*Occupations conducting interventions derived from scientific knowledge of mechanisms, structures, and context*" (2011, 17). "Professions such as psychiatry and social work often use dissimilar strategies and models. While the former administers mechanisms such as pharmaceutical drugs or talking therapies to transform the individual, social work uses another model, involving transformations of the individual's context so that his or her "normal" mechanisms again can begin to function. He employs his distinction between context and mechanism to understand such various types of interventions, saying that what constitutes a mechanism/intervention, "technological fix," or "magic bullet" in one model becomes context in the other, and vice versa (17). Social work professional knowledge tends to be context and structure oriented, whereas physicians and psychiatrists are more technologically and experimentally oriented.

Participatory Technology Development

The context orientation of social work knowledge perhaps stimulates a greater preoccupation with participatory strategies for service development and delivery. Policy innovations that involve organizations where knowledge workers are the agents of service delivery are likely to raise substantial and relatively enduring challenges to "understand the social relations of the workplace and their implications for systems design" (Hartswood et al. 2002, 10). Yet design developments typically exclude practitioners and are premised on the false assumption that the only relevant kinds of knowledge for such development are those of technology and management.

Hartswood and colleagues (2002) argue for a process of "co-realization," by which they mean that ICT professionals and the "users" of ICT systems should together create, implement, and continue to develop ICT systems. This argument has not been without its critics. Ross (2010), for example, argues that the users of technology in development must first be treated as objects rather than agents of social construction. However, Hartswood and colleagues suggest that the ICT professionals need to move beyond a narrow "engineering" mentality and physically move to work for long periods within the organization not just in the design phase but, crucially, in the subsequent implementation phase. It is in this latter phase, when people actually start to make use of a system for the first time, that there is the most opportunity for users to shape both their own ICT-enabled practice and the development of the system.

As users become "experienced" they develop new ways of using the system that in turn generate new ideas for its further development. Rather than users simply adapting themselves to the new system, co-realization stresses a change not only in the user but also in their use of the system, as a set of working practices evolve through use. Furthermore, we would argue that through this process users gain more general IT competences and become better able to judge inter alia what is possible and what is not, what is simple and what takes time (Hartswood et al. 2002, 24).

Hartswood and colleagues are placing greater stress on the process by which an information system is arrived at, not simply the system itself. An important question arises at this point, embodied in a distinction associated with Checkland and others between hard and soft systems (e.g., Checkland 1999; Checkland and Howell 1997).[5] Checkland brings to this an argument about systems as methodology based on three considerations about human problem situations.

1. There always will be different worldviews and different interests being pursued in human problem situations.
2. Worldviews change, and hence we are dealing with material that is not homogenous through time.
3. Human groups always include people who are trying to pursue purposeful action.

These assumptions regarding the complexity of human affairs took Checkland to argue that the way to tackle this complexity is through a methodology of learning systems. For him the system is the way we tackle the real world and not something thought to exist "out there." By contrast with this "soft" system, a "hard" system, for Checkland, is a system existing out in the world that may not work but that can be made to work better. The support that this gives for participatory ways of realizing technology is fairly obvious, as is the comparability with Giddens's arguments above and in chapter 7 regarding the relationship between the natural and the human sciences.

Checkland and Holwell regard the three decades of systems research at Lancaster as a form of action research (AR), but action research with a difference. Their concern is to present "an argument for an appropriate form of validation which, though it does not match the magic of replicability criterion in natural science, can sustain AR as a legitimate form of inquiry" (Checkland and Holwell 1998, 10). Applying the second point above, they argue that the classic methods of reductionism, replicability, and refutation cannot be applied in the human sciences and that in action research "the only certain object of research becomes the change process itself" (11). From a version of Giddens's double hermeneutic, they argue that the area of concern cannot be treated as an unchanging part of external reality and is not homogenous over time. Expressed differently, the act of inquiry applied to the area of concern may lead to learning about, and change within, the framework of ideas, the methodology, and the area of concern. Hence there must be a prior declaration of both theory and methodological process if there is to be a convincing truth value to claims from AR. "Without that declaration it is difficult to see how the outcome of AR can be more than anecdotal. Many literature accounts of AR leave the reader wondering . . . How is it to be distinguished from novel writing?" (14). In relation to the principle of replicability they conclude that "the aim in AR should be to enact a process based on a declared-in-advance methodology . . . in such a way that the process is *recoverable* by anyone interested in subjecting the research to critical scrutiny" (18). And so our discussion of technology has returned us to basic principles of epistemology.

A further relevant distinction has been made by Collins and Evans (2002, 2007) between the technical and political phases of decision making. These are not two separate decisions made at different times by different groups but are partly complementary. In the technical phase, although the primary focus is on testing propositions, that is not to say it is value free. It prioritizes a set of values about the importance of reason and evidence, associated with Mertonian ideals of science. Within what frameworks should such technical debates take place? The outcome of decision making is a strategy for action. This means that other factors enter in, not just expert knowledge. "Making this distinction does not, however, reintroduce the distinction between facts and values . . . because there is no assumption that the activities that take place within the technical phase are 'value-free' or 'neutral'" (Evans and Plows 2007, 834).

Technology Mediates Practice Mediates Technology

Paradoxically, technology always mediates practice. There are familiar and ubiquitous forms of such practice mediation that are often almost invisible because they are so taken for granted, from pencil and paper to cell phone, e-mail, two-way clinical observation mirrors, or the mutual recording required of some forms of intervention such as task-centered work. As we have seen already, some observers detect changes in the forms of knowledge in social work from the "social" to the "informational," expressed as a shift from a narrative to a database way of thinking. The thrust of policy innovation is to *separate* knowledge from the knowledge worker and make it something that can be manipulated independently. In this way the user of information technology is disciplined by it. "The 'information-age' is believed to have shifted our lives more towards the world of networks (virtual and actual) in which knowledge is defined by its utility and by its partializing, standardizing and universalizing functionality" (Pithouse et al. 2009, 603–604).

This is most commonly achieved through IT systems that seek to capture and codify knowledge. Insofar as this is true, there is a constant tension between knowledge as residing in people's heads and knowledge as "downloadable," transferable, and open to comparison with other cases. This is a helpful way of thinking about developments in social work, where there is without doubt a growing emphasis in Western social work on a standardization and formalization of service delivery. But it should not be overstated. Hence—and here is a second paradox—practice always mediates technology such that technology rarely if ever *replaces* existing practice. Nor is it easily imposed on practitioners. For example, rational, standardized technical forms are unlikely to replace or seriously subvert the exercise of professional discretion, judgment, and reflection. Evans and Harris have well reminded us of the irony that the "existence of rules is not inevitably the death-knell of discretion. Rather, by creating rules organizations create discretion" (2004, 993). Schön's (1983) illuminating comparison of superficially contrasting professions—none of which actually included social work—plausibly shows how professionals working in very different fields appear to exercise discretion and reflective practice even in cases such as architects and librarians, where technology is part and parcel of routine practice.

Taken together—technology mediates practice and practice mediates technology—we should see technology as itself a practice. It is something that practitioners accomplish, rather than it being a "thing." This is not the usual way practitioners see technology. For example, in a group of practitioners discussing their involvement in a practitioner research network, one member referred to an audio-transcription machine.[6]

> **Lesley:** Can I just add one thing—I am slightly worried about technology. I mean, I still don't know about where we get a thingy . . .
>
> **Lorna:** [Service Director] has one if you want one.
>
> **Lesley:** . . . and I am just confused.[7]

The work of a number of social work researchers has led them to question the conclusion that technology determines what social workers do. Research with social workers in Belgium concludes as follows: "The mutual opposition of technology and social services is limited. . . . Computerization is a part of structures, having influences and being influenced by human workers or human actions" (Laurent 2008, 383). It follows that "workers can also manifest strategic attitudes of resistance. . . . They always have the possibilities of playing with the rules" (383). Pithouse and colleagues reach a similar conclusion about IT programs in social work when they refer to "the potential for moderation and 'work-arounds' by staff in order to win some advantage" (Pithouse et al. 2009, 604). Shaw and Clayden (2009) refer to the presence of tactics to defend service users, whereby service users were deliberately not informed about technology innovations, online forms were adapted "on the hoof," and when the scheme was seen as requiring unrealistic expectations it was soft-peddled at the local level.

Example 4.1 gives a synopsis of the UK government's introduction of an approach to social work with children that places ICT at the heart of practice, and it presents some points from an evaluation of the system. The balance of the evaluation findings were, in this instance, critical, but they illustrate how the mutual mediation of practice and technology takes place in local services.

EXAMPLE 4.1

Mediation Between Technology and Practice: The Integrated Children's System

The Integrated Children's System (ICS) was an England and Wales government-led initiative from London and Cardiff, part of a wider package of developments for children's services, designed to promote effective services for children and families in England and Wales. It aims to help them do this "in a systematic manner, and to enable practitioners and managers to collect and use information systematically, efficiently and effectively" (Department for Children 2009).[8] The government laid out the key elements of ICS as:

1. An understanding of social work as consisting of assessment, planning, intervention, and review.
2. A set of data requirements providing common information from one locality to the other about children and families.
3. A set of "exemplar" formats for social work practitioners and other agencies, which form the basis for an e-social care record.

A national evaluation of the ICS drew five conclusions about the relationship between social work practice and technology (Shaw and Clayden 2009).

A. The ICS actively shaped practice. It:

1. Brought issues into focus, rendered social work visible, but served to partialize practice in a way that made it difficult to see the whole story.
2. Challenged practitioners to consider what counted as "important" or "serious" evidence and what counted as less "serious" or weighty.
3. Unhelpfully "fixed" the character of social work evidence.
4. "Pre-coded" some aspects of practice and left others "open ended."
5. Changed some language forms of social work to information terms.

B. ICS helped to disentangle and bring into focus and clarify, though not solve, issues that had been there previously but where existing recording and practice systems rendered them less visible.

C. The ICS made social work more visible, but it was a certain kind of visibility—a performance-culture visibility.

D. The ICS distanced the service user.

1. It was *overwhelming for service users* to cope with.
2. The *language* used in the forms was difficult to understand.
3. The *volume and density of information* required from the exemplars was seen as intrusive.

E. It led to new forms of discretion by practitioners as they found ways of working with the ICS at a local level.

Practitioners need not draw wholly negative conclusions from such research. Social workers should be neither optimists nor "doomsayers" regarding technology. But it does oblige us to think critically about each aspect of social work practice. Take recording as an example. An argument that, even at this distance in time, remains striking was made by the influential early sociologist Ernest Burgess at Chicago. The main case he develops (Burgess 1923) is about the potential but largely unrealized value of agency case records, especially regarding families. "What should social case records contain to be useful for sociological interpretation? They should contain what will render them valuable for social case work, that and no more. This answer will, I know, perplex and astonish many social workers and sociologists" (Burgess 1928, 524).

His specific proposal is enticing: "My proposal is actually quite simple and I think, entirely feasible and reasonable, in spite of the fact that I do not anticipate its immediate and general adoption. It is to enter into the case record statements made by all persons visited in nearly as humanly possible the language which they used" (Burgess 1927, 192). Quoting the prominent nineteenth-century British housing reformer Octavia Hill, he laments:

> Existing case records seldom, or never, picture people in the language of Octavia Hill, with their "passions, hopes, and history" or their "temptations," or "the little scheme they have made of their lives, or would make if they had encouragement." The characters in case records do not move, and act, and have their being as persons. They are depersonalized, they become Robots, or mere cases undifferentiated except by the recurring problems they present.
>
> (Burgess 1928, 526–527)

Why refer to writing nine decades old in the context of ICT? Because the more severe criticisms of records as made by social workers in the ICS study were to the effect that the possibility of a "narrative" of the kind Burgess would have recognized was lost in the detailed compartmentalizing that took place in the exemplar records. Burgess's remarks also confront social workers of today as of years-past Chicago with the question of whether recording technology moves too readily to professional accounts at the expense of hearing the voice of service users.

Taking a step back, Ulrich (2001a, 2001b) has offered one of the most extensive frameworks for theorizing information systems. He develops a discursive position on the definition, design, and development of information systems, by which he means "an argumentative method of clarifying disputed validity claims" (2001a, 59). In doing so he is applying general ideas and "rules" from pragmatism and shaping his understanding of the need for argument in the light of Habermas's theory of communicative action. He helpfully stresses the need to explore such questions empirically.

"The paper proposes a critical approach to information systems definition, design, and development (ISD) grounded in discourse theory, semiotics, practical philosophy and critical systems thinking . . . conceiving of information and knowledge . . . in terms of processes rather than objects" (2001a, 55, 56). His implicit idea is that "the value of "information" consists in contributing to knowledge for purposeful action and that the value of "knowledge" in turn consists in guiding people toward rational action in the sense of helping actors to choose courses of action with reason" (57). He sets out a series of nine steps—his "philosophical staircase"—by which one moves from information to action, and he does so by asking: What do we mean by information? How does information become knowledge? And how does knowledge relate to rational action?

He gathers his analysis into a practical way of "identifying the concerns of stakeholders; drawing conclusions concerning possible proposals for 'improvement'; and examining the validity claims embodied in such proposals in terms of their empirical and normative selectivity" (2001b, 102).

Questions, Questions

Having sought to "place" technology in relation to society, science, and professional practice, the final part of this chapter focuses on a series of more specific questions. How straightforward is it to interpret evidence from technology? How might a distinction between experiments and demonstrations inform social work thinking and practice? Has the expansion of technology enhanced the possibility of collaboration in social work science? What future issues face social work technology?

Interpreting Evidence

We noticed earlier Collins and Pinch's remarks that "scientific and technological debates seem to be much more simple and straightforward when viewed from a distance," yet ironically "the greater one's direct experience of a case, the less sure one is about what is right" (1998, 2, 3). Donahue interestingly captures the work of the Chicago Nobel laureate economist James Heckman. Heckman deals with "the daunting nature of uncovering complex causal relationships from nonexperimental data" (Donahue 2002, 24), developing from a concern that the quality of research had not improved. He advanced two reasons. First, "the technological and organizational factors . . . that have made high powered computing and large data sets widely available, coupled with the institutional efforts of the unscrupulous or untutored to offer empirical support for various policy measures, has enormously expanded the number of empirical studies of dubious quality" (28).

Second, Heckman judged there to be an intermediate category of empirical studies "that lies between the truly valuable and the utterly worthless." Here researchers have used more sophisticated techniques, but by adhering to these tools they have been lulled into believing they have found "true causal relationships, when in fact they have largely uncovered numerical noise" (28). The problem may not be exactly the same in other fields such as social work, but the general point is that the increased complexity of techniques "means that it will be more difficult for researchers and readers to have a feel for the data, and it will be more difficult to critique an ostensibly well-crafted study that isn't marred by one of the more obvious pitfalls" (20).

Rewards hence are more likely to go to those who master complex techniques rather than those who are alert to issues of data quality. We thus lose sight of the reality that "the truth is incredibly hard to uncover and it cannot be assured simply by following a state of the art statistical protocol" (29). Any successful empirical research agenda needs not only knowledge of methods and techniques but also "less concretely identifiable skills of creativity, sound and tutored judgement, relentless effort to find and create more and better data, and exacting care of all phases of execution" (29).

Experiments and Demonstrations

The difficulty of interpreting evidence from technology is narrated in a series of fascinating examples by Collins and Pinch, regarding, for example, whether Patriot missiles shot down Iraqi Scuds in the first Iraq war, the debate about assigning blame for the *Challenger* explosion in 1986, and safety testing of nuclear fuel flasks and antimisting kerosene in plane crashes.

The last two focused on train crashes to simulate nuclear flask safety and a U.S. plane crash to test "safe" kerosene. This is about two public demonstrations proceeding from different logics to lab testing. Demonstrations are uncontroversial. "One does not do experiments on the uncontroversial, one engages in demonstrations" (Collins and Pinch 1998, 62). The demonstration "shows" and "displays"; the experiment "tries." Trying occurs in relatively private spaces; demonstrations take place in relatively public spaces. "Demonstrations are designed to educate and convince once the exploration has been done and the discoveries have been made, confirmed and universally agreed. . . . Indeed, the work of being a good demonstrator is not a matter of finding out unknown things, but of arranging a convincing performance" (63). "It is very important that demonstration and display on the one hand, and experiment on the other are not mistaken for one another" (64).

If this seems a barely relevant excursus, it is valuably developed by Dearing. He offers a useful distinction between experimental and exemplary demonstrations, which is directly parallel to Collins and Pinch's distinction between experiment and demonstration. An *experimental* demonstration is "a field test carried out for the purpose of assessing external validity of an intervention by varying the setting, the participants, resource availability, implementation protocol, or the methods by which outcomes are measured" (Dearing 2009, 511). "The purpose of an experimental demonstration is data collection. . . . 'Does this model work under real-world conditions?' " (511). He calls this a "prediffusion activity," and it is crucial to knowing if an innovation should be diffused.

An *exemplary* demonstration is "a persuasive event calculated to influence adoption decisions and thus increase the likelihood of diffusion" (511). The purpose is not merely dissemination but "to showcase an intervention in a convincing manner. . . . Exemplary demonstrations increase the likelihood of diffusion partly by making a costly, worrisome and complex intervention more understandable through visibility of its processes and observability of its outcomes" (510).

He laments that "lack of clarity about the purposes of demonstration is a frequent culprit in the nondifffusion of effective interventions. . . . A disconfirmed hypothesis that leads to a design improvement is a positive result in an experimental demonstration; in an exemplary demonstration, such an outcome is noise that will lead to perceptions of higher, not lower, uncertainty among potential adopters" (511).

EXAMPLE 4.2
Experimental and Exemplary Demonstrations

The Integrated Children's System (ICS) study referred to above received heavy investment by the governments in London and Cardiff. It was one central government response to a much-publicized child death (Victoria Climbié). Two extensive studies were commissioned. One was undertaken by the team of academics who also had been the lead developers of the electronic formats ("exemplars") at the heart of the initiative. The second study was commissioned from a separate team to undertake a formative evaluation of the coherence, understandability, and acceptability of the ICS. The focus of each brief was intended to be complementary.

The final report of the second study identified a series of deep-seated problems with the ICS and made a series of specific linked recommendations. The criticisms dealt with the prescriptive nature of the system; the inflexibility and standardization within the exemplars; ways in which the length of the records, their format, the language, and the underlying assumptions made it difficult to promote user involvement; the unrealistic or inappropriate time scales; serious difficulties in interagency communication and information sharing, in developing interfaces with other agency databases, and consequently of "early warning" of risk to children; the unclear balance between central government lead and local autonomy; and ways in which the arrangements for commissioning ICS package providers was a cause of serious difficulties.

The first commissioned study team reported and reached much less critical conclusions.

The central government department declined to accept the report of the second team. Written questions were twice asked in the House of Commons during 2008 as to the nonpublication of that report. Extensive press coverage was afforded during 2009 in relation to the team's conclusion that "the evidence from the evaluation suggests that the ICS has yet to demonstrate the degree to which and how it is fit for purpose."

Sustained dissent of this kind usually has multiple causes, but part of the issue appeared to be that the government saw it as a demonstration.

Tentatively, I suggest the first study team also saw it as a demonstration, whereas the second team saw it as an evaluation, test, and, in the sense Dearing intends, an experiment. For the demonstration-oriented team, the ICS was not controversial. They were seeking to exemplify the policy initiative and thus held a general confirmatory approach. For the experimental test team it was controversial, and members were open to—and indeed searching for—falsifying evidence. Hence the second team's conclusion that the ICS was not fit for its purpose was almost impossible for government officials to accept because it was a demonstration, a model, with a modest development role to iron out implementation issues.

Source: https://www.gov.uk/government/uploads/system/uploads/attachment_data/file/273183/5730.pdf; Shaw et al. (2009).

Do Technological Developments Improve Collaboration in Social Work Science?

Unlike laboratory-based sciences, there has been little attention to the character of collaboration between social work scientists or to if and in what ways research networks exist. This second absence was touched on in chapter 2. "Today, dispersed collaborations are more feasible because communication technologies allow scientists to exchange news, data, reports, equipment, instruments, and other resources" (Cummings and Kiesler 2005, 704). In principle this may seem to enrich the potential for "distributed intelligence" and for cross-disciplinary work. "An important claim favoring multidisciplinary collaborations is that they promote innovation" (704). I cite this study because it seems to indicate that "despite widespread excitement about dispersed collaboration reflected in terms like 'virtual team,' 'eScience,' and 'cyberinfrastructure,' there appear to remain a number of challenges that scientists encounter when they work across organizational boundaries." Thus "technology did not overcome distance" (717). For example, on the use of e-mail they interestingly say: "Our impression from the workshop was that email was used a great deal, but that it failed to help people coordinate project work across many investigators located at different places. Using email sometimes encouraged too much task decomposition and too little intra-project sharing and learning" (718).

Futures

Schuerman suggested that "one of the things that distinguishes social work is that we try to make use of the creative tension among passion, analysis and technology" (1987, 17). In so doing we may anticipate what is feasible and what is not. There are several meanings when people say something "cannot" be done. We might express this as the meaning of the impossible. Collins (2010) suggests that two of these are:

- "Cannot" as technological impossibility: something that would need the invention of a new principle that as our present knowledge stands cannot be foreseen
- "Cannot" as technological competence—for example, "we cannot translate the Rosetta Stone"

These "cannots" vary historically. They are true at one point but not at another. I recall the first moment, now twenty years ago, when I encountered two colleagues cowriting a book, one in the United States and one in England, and my sense of something new when I learned that one would revise a chapter and e-mail it to the other at the end of the day, who would rise the next day and work on the next stage of revisions, which had arrived while he or she slept. People will disagree whether something is an instance of technological impossibility or competence. This is an example of a potential disagreement in social work science. In the next chapter we look fully at the nature and significance of such disputes and disagreements.

Taking It Further

Reading

Checkland, P., and S. Holwell. 1997. *Information, Systems, and Information Systems: Making Sense of the Field.* Chichester: John Wiley and Sons.

Checkland, P., and S. Holwell. 1998. "Action Research: Its Nature and Validity." *Systemic Practice and Action Research* 11(1): 9–21.

Collins, H., and T. Pinch. 1998. *The Golem at Large: What You Should Know About Technology.* Cambridge: Cambridge University Press.

Forster, E. M. 1954. *Collected Short Stories*. Harmondsworth: Penguin.
Parton, N. 2008. "Changes in the Form of Knowledge in Social Work: From the 'Social' to the 'Informational'?" *British Journal of Social Work* 38(2): 253–269.
Ulrich, W. 2001a. "A Philosophical Staircase for Information Systems Definition, Design, and Development: A Discursive Approach to Reflective Practice in ISD (Part 1)." *Journal of Information Technology Theory and Application* 3(3): 55–84.
Ulrich, W. 2001b. "Critically Systemic Discourse: A Discursive Approach to Reflective Practice in ISD (Part 2)." *Journal of Information Technology Theory and Application* 3(3): 85–106.

Forster's short story "The Machine Stops" still has the power to entice and disturb. Parton's article is one of most complete statements of an important critical strand in writing about technology. But while I raise some substantial criticisms of how technology has been harnessed to a modernizing agenda, my own research has taken me to a position where I am not a doomsayer. Checkland's work on soft systems methodology has been influential and much cited. A good introduction that connects to themes with which social work has an existing familiarity is Checkland and Holwell (1998). Collins and Pinch (1998) have a delightful series of case studies that I draw on here and elsewhere in the book. Collins is sometimes earmarked as a relativist product of the "strong school" in the sociology of knowledge, but his work here, and also where discussed in chapter 7, provides slim grounds for that assessment. A more substantial source is Checkland and Holwell (1997), and, for those working in the information systems field, the papers by Ulrich add indispensable weight.

Task: Social Workers as Knowledge Workers

The idea that social workers are "knowledge workers"—a term coined by Peter Drucker—is helpful in developing a reflective understanding of the implications of how ICT is embedded in everyday practice and in connecting developments in social work to those in other comparable disciplines and professions. For a useful discussion of the general idea, the Wikipedia entry is a fairly good overview (http://en.wikipedia.org/wiki/Knowledge_worker). For a critical review of debates, see Darr and Warhurst (2008).

How might this idea inform how we think about the relationship between science, technology, and social work?

[5]

The Social Work Science Community

Controversies and Cooperation

I have bracketed a major question that can no longer be ignored. In what sense is social work a community in relation to its science work? How is social work science socially organized? Does it make sense to see social work nationally or even internationally as, in this regard, a shared collective enterprise? In the first part of the chapter I explore how far it makes good sense to see social work as engaged in science work collaboration. In addition, how far can we detect networks or perhaps schools? At the instrumental level, what mechanisms seem to exist for ordering the community?

Having explored this territory, the question arises whether this places undue weight on the collective rather than the individual within social work science. I will look at the idea of "inventions" in this connection, taking task-centered practice as my main example.

Despite the implicit diversity in such arguments, it may seem that it is all rather too consensual. To get inside this question we will consider two related questions. First, how far does it make sense to see social work as fissured by interests? Second, is social work and its science activity marked by controversies? The key words *interest* and *controversy* are given strong meanings in this discussion.

Finally, accepting that at some level there are differences and disagreements in social work on science issues, are there beneficial ways in which disagreements reciprocally should be explored and perhaps resolved or at least taken forward?

Collaboration

Whether we consider patterns of collaboration, research networks, scientific schools, or scientific inventions,

> we have to place science in the context of work if we are to understand why it is that scientists will produce knowledge in the way they do. We need to have a grasp of the inward conditions that motivate scientists, the ways in which scientists themselves are making sense of their project, and . . . we must have some understanding of the external conditions of science that pattern and structure the vocation of scientists. The interplay of these three factors is what gives science and scientific institutions their character.
>
> (Erickson 2002, 53)

The desirability of collaborative endeavor may seem to be a given, yet "tensions and paradoxes are essential features of collaboration, even within established, co-located research groups, so the mere occurrence of face-to-face interaction does not insure that understanding and solidarity will result" (Hackett 2005, 668). Hackett expresses these tensions as "openness and secrecy, cooperation and competition, priority and patience, dirigisme and autonomy, craft and articulation work, role conflicts, and risks of various sorts" (670).

What are we talking about when we speak of collaboration or of a social work science community? Who belongs? Is membership based on who you are or what you do, your sense of identity, or where you work? How do we know if particular social work program faculty, social work practitioners, contract researchers, or service users are part of such a community? Hackett suggests that what collaboration is and why scientists should collaborate are "deceptively simple questions" that "have elicited complicated and qualified answers" (668). He makes a series of valuable distinctions. These include:

> *extent*, measured as a distribution over substantive, social, or geographic space, or over time; *intensity*, measured as the frequency or significance of interaction among persons, places, or units of time; *substance*, or the aims and content of collaborative work, which now include producing fundamental knowledge, developing technologies, guiding decisions,

> making things, training, and bonding; *heterogeneity*, or the variety of participants, purposes, languages (ethnic, national, disciplinary, sectoral), and modalities of collaboration (face-to-face, electronically mediated in various ways, and episodic); *velocity*, or the rate at which results are produced, analyzed, interpreted, and published; *formality*, ranging from contractual arrangements among nations or organizations to handshake agreements and unstated understandings among friends and acquaintances.
>
> (669, italics added)

We should add to this that "scientists from different disciplines have usually trained in different departments, have had different advisors, publish in different journals, and attend different conferences. Their social bonds are likely to be comparatively weak" (Cummings and Kiesler 2005, 704). A glance at the publication records of established faculty will show that this is the case even with cognate disciplines such as social work, psychology, sociology, and policy studies.

National boundaries, even where first languages are shared, are also likely to shape the extent to which forms of collaboration occur. Take, for example, the extent to which social work scholars write collaboratively. A crude but useful indicator of this is the proportion of journal papers that are sole or joint authored.[1] A comparison of a leading British with a leading American journal suggests that U.S. scholars are more likely to publish jointly than those from other countries (and that the U.S. journal is more likely to publish coauthored articles than the British journal).[2] (See tables 5.1 and 5.2.) National characteristics and differences are also significant for collaboration if we take into account the breathtaking extent to which U.S. scholars rely solely on reading work by other U.S. writers (Shaw 2014e).

TABLE 5.1

Sole and joint authorship of social work articles: U.S. and non-U.S. writers

	SOLE AUTHORED	JOINT AUTHORED		ROW TOTALS
	N	N	%	
U.S. writers	13	71	85	84
Non-U.S. writers	30	46	61	76

TABLE 5.2
Sole and joint authorship of social work articles: *SWR* and *BJSW*

	SOLE AUTHORED	JOINT AUTHORED		ROW TOTALS
	N	N	%	
Social Work Research	13	67	84	80
British Journal of Social Work	31	49	61	80

I am not aware of any careful study of collaboration between faculty in different university social work schools in the United States, but the general assumptions of interuniversity collaboration were tested by Cummings and Kiesler, whose work on the implications of technology for collaboration we touched on in chapter 4. Taking principal investigators (PIs) as their indicator, their several conclusions were that

> to a statistically significant degree, more PI universities involved in a project predicted fewer coordination mechanisms used in that project. More PI universities on a project predicted a lower level of faculty, post-doctoral, and graduate student direct supervision. . . .
>
> The results also show that, with more universities involved, the pattern of coordination mechanisms changed. PIs were more likely to hold a conference or workshop and to work on the project at a conference or a workshop. . . .
>
> The analyses taken as a whole suggest that distance and organizational boundaries interfered with those coordination mechanisms that involve frequent, spontaneous conversation and problem-solving.
>
> (Cummings and Kiesler 2005, 710–711)

They proceeded to say: "Having more PI universities on a project was significantly negatively associated with the generation of new ideas and knowledge" (711). On the benefits of collaboration Hackett concludes: "Collaborations may generate papers that could not otherwise be written, and they may bestow a spectrum of advantages on participating scientists, groups, and organizations, but disproportionate productivity does not appear to be one of the benefits" (Hackett 2005, 669). He refers to a study in the journal issue he is introducing that seemed to show that collaboration in Ghana, Kenya, and Kerala had negligible, even negative, effects on productivity—the social context in those nations and states appears to undermine the potential for collaboration. Not to sound unduly pessimistic, this offers an agenda for social work, and "theories of innovation and social networks could benefit from further investigations of how weak ties change into strong ties during the collaboration process" (Cummings and Kiesler 2005, 717–718). "Currently we have no theory of the 'ideal' level of collaboration in science," and "in future research we should examine how different kinds of science use different forms of collaboration" (718).

Science Networks in Social Work

We discovered in chapter 2 that a search on "networks" in social work suggests tentatively that most of the articles that discuss networks are about action research or practitioner research, not mainstream social work research. In addition, the articles are mainly about intentionally formed networks—sometimes with a capital *N*. I found no extended discussion of the nature of social work research networks and, still more clearly, almost no empirical work on the form, organization, or significance of such networks.[3]

Once more we find that terms that seem straightforward easily prove troublesome. Rather than offer a definition of "network," example 5.1 quotes from four people who are members of a small international network that meets somewhere in the world annually. It has no formal constitution, no home site, no funding, and no external accountabilities. It was originally constituted about fifteen years before this study, with a membership of social work evaluation centers linked to some degree to government programs, evolving gradually into a more individual membership. The study drew on a mix of survey, network mapping, interviews, documents, and participant observation.[4] The study suggests that the array of themes includes:

1. Origins and "creation stories"
2. The nature of networks
3. Being a member
4. How they work

5. Influence on the outside "world"
6. Developments, changes, and trends in the network
7. Transitions, challenges, and futures

The extracts in example 5.1 touch on the nature of this network and how it "works."

EXAMPLE 5.1
A Working Network

Matthäus: I think that every member has a special role and this in various aspects. To give another example, I often see Alexis in the role of a critic who casts a different light on the debate, by fundamentally faceted brief interventions. Others, such as, for example, Luca, have a more integrative role, connecting topics and people.

While I talk, I realize that roles in the network are not that fixed. The just-mentioned roles of criticism and integration, for instance, can actually be taken by each member of the network. By the way, I would regard role flexibility as a characteristic of well-functioning networks, or even general social groups. I see [Network] as a very well-functioning network . . . except perhaps with an underrepresented management function. Some more management and organization could be good for the network . . . *could* be good. I am not sure about that because on the other hand the nonhierarchical and informal structure of the network contributed to its success.

Luca: I enjoy the fact that it is possible that scientists with quite different backgrounds and quite different theoretical and arithmetical orientations meet and get into a kind of exchange that is really driven by wanting to tell how the other person sees the world, or the questions that are . . . completely free exchange, the feeling that I can say whatever I want and I don't have to care too much if it's appropriate or if it's well formulated or if it might endanger my position or whatever, it's, it's, this is the real pressures experienced, I, I always miss when I'm not joining, because there's no other network or social group in my professional world that comes into the same thing.

Alexis: . . . then I have had these discussions with Logan and John, and Luca and many others here in the network, who have been doing more the things I would like to do, so I've learned a lot with these, about these discussions, from these discussions, and it has been an important part in my development as a researcher, and I hope in the future it is going to be like that, and also the fact that I, I had the opportunity to present my own ideas there in a free way, without any kind of censorship or hindrance, I have found very important for myself, to test ideas openly without punishment . . .

. . . it is really rewarding to work together with somebody else, I know this because I have been working alone for many years, and I know the agony which is connected to this lone work. But maybe I should say that you can still be individual when you are working together with somebody else; it's only part of the work.

Maja: I think it is different from most other networks, partly because of the focus on the research methodological issues, partly because of the, yeah, more pluralistic discussions and partly because of, that it is a more democratic network, where a lot of people over time take responsibility for arranging meetings and editing books and these kind of things, so I think it's, it's unique, this network, and it has been a network over quite a long time, and lots of networks are good for a few years and then they fade away.

Diffusion of Innovation

Understanding the diffusion of innovation is relevant because networks will be linked to aims of diffusing ideas; also, the way knowledge diffuses in organizations may help us understand how research networks operate.

Dearing refers to diffusion as involving "tricks of the trade"—hence pointing to both informal and formal diffusion processes. In any one sector "the state of the science (what researchers collectively know) and the state of the art (what practitioners collectively do) coexist more or less autonomously, each realm of activity having little effect on the other" (Dearing 2009, 504). He distinguishes diffusion and dissemination and helpfully says that "dissemination embeds the objectives of both external validity, the replication of positive effects across dissimilar settings and conditions, and scale-up, the replication of positive effects across similar settings and conditions" (504).

> Diffusion occurs through a combination of (a) the need for individuals to reduce personal uncertainty when presented with new information, and (b) the need for individuals to respond to their perceptions of what

> specific credible others are thinking and doing, and (c) to general felt social pressure to do as others have done. Uncertainty in response to an innovation typically leads to a search for information and, if the potential adopter believes the innovation to be interesting and with the potential for benefits, a search for evaluative judgments of trusted and respected others (informal opinion leaders).
>
> (506)

Dearing laments that in the United States, "not one American school of social work has translation, diffusion, or dissemination of effective practices, programs, or policies as its *forte*. Not one" (504).

Social Network Theory

Further clarification of what we mean when we speak of a network is yielded from social network theory. An example of network analysis within organizational research can be found in a substantial national evaluation of a UK government initiative, the Children's Fund (Edwards et al. 2006). The Children's Fund was launched in 2001 as part of the UK government's commitment to tackle disadvantage among children and young people. The program aimed to identify at an early stage children and young people at risk of social exclusion and to make sure they received help and support to achieve their potential. It encouraged voluntary organizations and community and faith groups to work in partnership with local statutory agencies, children, young people, and their families, to deliver high-quality preventative services to meet the needs of communities.[5]

The authors ask: how does knowledge "move around," upstream and downstream, within an organization? How does knowledge move between *levels,* between *strategy and practice,* and between different "*spaces*" or "fields" within an organization—for example, between child protection teams and teams working with children with a disability? They suggest that three types of informal networks among practitioners were evident:

- New trails trodden for the first time between individual practitioners who recognized the benefits of a collaborative response to the social exclusion of a child. They describe these as "light etchings or traces on a local landscape" (193).

- Networks that built on old networks and relationships but where there was evidence of the impact of the preventative intentions of the Children's Fund.
- Old established networks that were continued or resuscitated and where there was little evidence of the impact of the Children's Fund.

(Edwards et al. 2006, 193)

Hudson (2004) may be unduly sanguine when he suggests:

> It is within networks that . . . the "entangling strings" of reputation, friendship, inter-dependence and altruism become an integral part of the relationship. Accordingly the information communicated is "thicker" than that in the market and "freer" than that in a hierarchy—a prerequisite to addressing issues with uncertain solutions, and a factor likely to encourage mutual learning and innovation.
>
> (79)

But social work research on contexts and contextualization does demand "socially shared understandings of the normative contours of 'proper places' which shape the way people respond to the everyday lived reality of places" (Popay et al. 2003, 55).

"Schools" in Social Work?

Disciplines are often thought to develop through the emergence of "schools" of research or thought. But confidence of definition is as hazardous here as for the disciplines themselves. Are schools relatively uncommon, or does everyone belong to one? Does someone know if they belong to a school? There are doubtless numerous affinities and propensities within social work science—those who are committed to and largely confident in the value of empirical evidence, those who hold a constructivist stance, or those who are driven by a political or justice-based agenda. There are also more clearly defined communities of interest: the disability movements, aficionados of the Cochrane or Campbell collaborations, and so forth.

Are these "schools"? In social work this term has, as often as not, been about practice "schools," for example, functionalist or psychodynamic. Edith Abbott referred interestingly to "what is known among social workers as 'The Chicago School'" (1931, 132). Her "what is known as" may suggest a meaning of "school" close to the sense discussed here, albeit it would be an outlier in the context of her frequent reference to "School" in relation to the Graduate School of Social Service Administration.

"Being influenced by a tendency is not the same as belonging to a school" (Platt 1996, 236–237). Platt suggests that "school" in this context is used in three ways:

- School of current social membership
- School of retrospective identification
- School of imputation

This is helpful for how it distinguishes categories that refer to the sphere of ideas from those that refer to social relationships. Yet the concept and its reference points remain elusive. It is unlikely that temporal and geographical centers of research work capture even the best work.

Perhaps the most frequently named school in the social sciences is the Chicago School (of sociology), or as Howard Becker has called it, "*The Chicago School, So Called.*" I refer to it for how it invites ways of thinking about "school" claims in social work. He speaks of

> a school in the sense that historians of thought speak of a school, or what French intellectuals sometimes refer to as a "chapelle" (a chapel). In the structure of such a school, one person's thought is usually seen as central. When sociologists speak of a Durkheimian school, they mean to indicate, and with good reason, that everything connected with that school of thought was of a piece. The theory was and is consistent and coherent. The theory informs the research done in its name. The followers or acolytes preserve the founder's memory, embellish the theory and its associated body of thought, and further its fortunes, correcting errors and inconsistencies in the master theory and doing work that exemplifies its vision.
>
> (Becker 1999, 4)

But he insists that Chicago sociology was never a school in that sense.

> "Chicago" was . . . a school of activity, the activity being the training of more sociologists, and the awarding of degrees, and the maintenance of

> a reputation within and beyond the university. . . . The moral of today's story is that "Chicago" was never the unified chapel of the origin myth, a unified school of thought. It was, instead, a vigorous and energetic school of activity, a group of sociologists who collaborated in the day-to-day work of making sociology in an American university and did that very well. But we cannot make an inferential jump from that pragmatic collaboration to a "tradition," a coherent body of theory. The real legacy of Chicago is the mixture of things that characterized the school of activity at every period: open, whether through choice or necessity, to a variety of ways of doing sociology, eclectic because circumstances pushed it to be.
>
> (10)

Jennifer Platt's (1996) comments on Stuart Chapin at Minnesota are very interesting. His work, she suggests, was both meritorious (methodologically sophisticated) and fairly extensive, but he never became key in sociology history. Of equal interest is that he never became at all known in social work history. Yet he wrote work that was acutely relevant to social work (Chapin 1920, 1947; Chapin and Queen 1937).[6] Indeed, he was probably "caught" from both sides in that his social work focus was a reason why sociologists did not recognize him, along with the fact that he was not at either Chicago or Columbia, then the corporate *doyens* of social work research.

Ordering the Community

However we depict the nature of interrelationships within the social work science community, there are a number of mechanisms and procedures within social work broadly defined that embody the efforts of various interests to promote and assess well-grounded social work. Peer review, citation metrics, journal rankings, and national research assessment exercises are among such mechanisms. Simply to list them is sufficient to raise "Yes, but . . . " responses in readers' minds. Take, for example, a ranking of social work journals to hand at the time of writing.[7] Citation indices are calculated by dividing the number of citations of recently published items by the number of recent items (see table 5.3).

What sense one may make of table 5.3 is something of a mystery. American journals are to the fore, followed by British ones and a smattering of others. Scattered thoughts invite question upon question. Does this represent

TABLE 5.3
Social work journal citation report rankings

RANK	JOURNAL	2 YR IF
1	*Trauma, Violence, and Abuse*	2.939
2	*Child Maltreatment*	2.706
3	*Child Abuse and Neglect*	2.135
4	*American Journal of Community Psychology*	1.968
5	*Journal of Social Policy*	1.632
6	*American Journal of Orthopsychiatry*	1.504
7	*British Journal of Social Work*	1.162
8	*Health and Social Care in the Community*	1.151
9	*Social Policy and Administration*	1.143
10	*Revista de Cercetare si Interventie Sociala*	1.141
11	*Children and Youth Services Review*	1.046
12	*International Journal of Social Welfare*	1.021
13	*Research on Social Work Practice*	0.905
14	*Health and Social Work*	0.895
15	*Social Work*	0.877
16	*Family Relations*	0.862
17	*Qualitative Social Work*	0.836
18	*Journal of Community Psychology*	0.832
19	*Child and Family Social Work*	0.824
20	*Social Service Review*	0.791
21	*Child Abuse Review*	0.787
22	*Journal of Social Work*	0.709
23	*Children and Society*	0.684
24	*Social Work in Health Care*	0.66
25	*Australian Social Work*	0.591
26	*Journal of Social Work Practice*	0.586
27	*Social Work Research*	0.535
28	*Australian Journal of Guidance and Counselling*	0.5
29	*Clinical Social Work Journal*	0.488
30	*Administration in Social Work*	0.462
31	*International Social Work*	0.451
32	*Smith College Studies in Social Work*	0.441
33	*Journal of Social Work Education*	0.439
34	*European Journal of Social Work*	0.352
35	*Social Work in Public Health*	0.333
36	*Journal of Social Service Research*	0.309
37	*Families in Society*	0.298
38	*Affilia*	0.18
39	*Asia Pacific Journal of Social Work and Development*	0.152

a measure of good work? Where are the great majority of journals in languages other than English? What do the figures mean? Assuming they measure something about reputation, do we believe the relative positions of different journals? What do we think of the way highly regarded journal publishers gently encourage editors to commission articles that seem to have higher citation rates? If all publicity is good publicity, are we content that articles cited only to criticize them severely have equal weight with those that are cited approvingly? Given how far U.S. social work scholars limit their citations to other U.S. writers, how do we regard the collective self-referencing character of journal rankings? Blyth and colleagues (2010) have reviewed critically the nature and significance of journal metrics. They remark, for example, that

> There can be no automatic assumption that citation is equivalent to either approval or value. Citation per se may, for example, indicate that a paper (a) is controversial/bad; or (b) that other authors refute its contents. . . . Similarly there can be no automatic assumption that citation means the original paper has even been read by the person citing it! Simkin and Roychowdhury . . . report a method for estimating what percentage of those who cited a paper had actually read it and conclude that about 20% only of citers read the original. Lawrence . . . reviewed the 48 citations of a paper of which he was co-author . . . and claims that of these, only eight were appropriate.
>
> (132–133)

The remarks of Blyth and his colleagues move from citations combined into impact factors to index a journal's standing to citations as measures of individual achievement, but once again little is disclosed. Compared to other fields, social work citations as proxies for classic reputations are very low on visibility. The main citations are of textbooks, and the work of some who on the basis of fairly widely held judgment within scholarly networks count as leading research scholars is lowly cited. Readers may select someone they consider the major social work research scholar of a generation ago and take one of the standard citation resources, only to find that their citation levels are only a fraction of those for textbooks by well-known writers for the student market.

Pressures on social work scholars to publish or perish and to have their work published in the most highly regarded journals means in turn that peer review also has come under attack. In 2014 the journals *Nature* and *British Medical Journal* were both plunged into controversy and obliged to withdraw and retract articles on stem-cell research and the use of statins.

Despite these criticisms, there is a seductive attraction to citation figures and impact factors. Past editors of a journal implied this attraction in the following way:

> In the UK at the moment officially the impact of the journal doesn't necessarily matter, but I find that a bit hard to believe really . . . [so] much as we are skeptical about impact factor and league tables (we wrote a paper about it), at the end of the day they are the rules of the game and if people want to prosper in the game they have to play by the rules, whether they think they're crazy or not.

In Thomas Kuhn's language, all these manifestations can be viewed as part of the regulatory if sometimes brittle order of "normal science," assessing work against cultural and academic norms and standards set by those with the power to do so.

Social Work Inventions

Having explored collaboration, networks, schools, and community ordering mechanisms, the question arises whether this places undue weight on the collective rather than the individual within social work science. The appearance and rise of task-centered practice illustrates how this issue may become central. William Reid has been called the inventor of task-centered work. The tribute to him on the award of his distinguished professorship in 1998 recorded: "He is principally known as the inventor of the task-centered approach, widely recognized as pioneering a new method and philosophy of practice for social work, a field that had been steeped in long-term psychoanalytic practice prior to his research and writings."[8]

James Lovelock, the famous chemist and writer on the Gaia principle, opined that "invention is quite easy. . . . Most people are inventors if they only knew it. What is really clever is thinking of the need" (Wolpert and Richards 1997, 73). The place of "inventions" and "discoveries" was introduced in chapter 3. Task-centered practice has often been presented as the triumph of a visionary lone hero standing in an honorable succession and with the support of a few key likeminded people. Videka and Blackburn (2010) offer a good example of this way of painting the scene. For them, Reid "shaped the scholarly agenda and the practice of social work throughout his thirty-five year career" (183). His work "transformed the profession from one that had little investment and stake in empirical knowledge development for practice to one that is deeply invested in scientifically based practice approaches." He was "a visionary of his day" who saw that his approach "could revolutionize the practice of social work" (183). Indeed, "task-centered practice revolutionized social work practice, transforming it from tradition and ideology-based approaches to empirically based practice" (191) and in doing so "hastened the shift to new practice paradigms" (Fortune 2012, 31).

Fortune and colleagues, in a circumspect version of the lone-hero narrative, describe Reid as "a solo researcher with few grants who nevertheless systematically constructed, tested, adapted, replicated and disseminated an intervention model" (Fortune, McCallion, and Briar-Lawson 2010, 284). They note the growth of research centers but seem to plead the relative value of the solo model in saying that despite the proliferation of centers and training opportunities for advanced research methods, "it is unclear how many centers and studies focus on intervention" (285), and "Reid's solo research model continues to be relevant" (291). In research centers, because of short-term and funder-priority-led priorities, "it is difficult to establish a systematic trajectory of model-building research" (285).

Reid's individuality is captured in Fortune's recall of his way of working:

> Bill had a quirky way of thinking that was his greatest gift, I think. Although he was (obviously) a linear problem-solver, he also had a lateral way of thinking that enabled him to take something mundane and turn it on its head to create something brilliantly new. He would stare at something, or listen to tape recordings, and suddenly come up with an insight about the something while others were already assuming the "obvious conclusions" (which were not at all interesting). I don't really know how to describe it. He had his students tape-record their interviews (later, video-tape) and he was always listening [to] or watching them. His office at home and school were set up like viewing stations.[9]

Reid and Shyne were more circumspect than some later writers in claiming ex nihilo invention. There had been for some time counteracting arguments as to whether the merits of long-term casework had been demonstrated or had been oversold. The advocates of the functional school of casework had held that a fixed limit on the duration of intervention was an essential part of the helping process, as had crisis theorists. Others have laid still greater stress on the influence of community-level influences. Several writers and contemporaries have referred to the cluster of social work doctoral students at Columbia University in the early 1960s who had mutually reinforcing interests and who made significant subsequent contributions to the empirical practice movement in the United States. Ed Mullen was a member of this group and recollects how

> Regarding the invention of task centered . . . Bill and I were working very closely together at CSS[10] when he and Ann Shyne designed the casework methods project. I used his data as it was being collected for my dissertation process analysis and was on the scene with analyses as the findings were coming in. We were together trying to figure out what about short term treatment made the difference and out of that emerged the ingredients of the task centered idea.[11]

In Reid's case, the interests of the Community Service Society and the movement to and from Columbia and Chicago also contributed (Fortune 2012). There were important elements of continuity with the past. Reid's dissertation chair had been Florence Hollis, and it is possible that his engagement with her typology of social work interventions, which first appeared in book form in 1964, helped generate his interest in process research. Finally, remembering that history gets told by the winners, it also is possible that accounts of inventions give less weight to the roles of those outside the academy. There are intriguing references in some published accounts (e.g., Reid and Shyne 1969) and in the memories of those who were directly involved to the part played by practitioners—something that always is likely to fade into the background.[12]

In the light of these counterpoised narratives, how might we better understand the emergence of task-centered practice and, by implication, other "inventions"? In a neglected part of Michael Polanyi's (1966) argument regarding tacit knowledge he asks how originality and innovation take place, and he answers that it is by surprising confirmations and by granting the highest encouragement to dissent. This is because "the metaphysical beliefs of scientists necessarily assure discipline and foster originality in science" (70).

He sees the scientific community operating through a "*principle of mutual control*" (72) through "the simple fact that scientists keep watch over each other. Each scientist is both subject to criticism and encouraged by their appreciation of him." This produces "*chains of overlapping neighbourhoods*" that in turn yield a "mediated consensus" (73). When people newly join scientific communities they trust the tradition while at the same time claiming an independent position from which they may reinterpret this tradition. "Scientific originality springs from the scientific tradition and at the same time supersedes it" (75). How things appear to the participants is less straightforward. Looking backward, new ideas can be seen as emerging from the pathways of science and hence predetermined. They "represent the possibilities that were previously hidden and dimly anticipated in a problem." Yet looking forward, to the scientist "the act of discovery appears personal and indeterminate" (75).

Polanyi's counterpoising of tradition and change, predictability and indeterminacy, community and independence, offer a congenial, if somewhat sanguine, perspective from which to understand task-centered practice. Consistent with his image of scientific work, it is possible to develop two somewhat counteracting narratives on the invention of task-centered practice—the personal and indeterminate work largely of the key scholar and the narrative that explains the invention within a social context of diverse interacting interests. Each narrative conveys a different understanding of the community of science.

A Successful Invention

Why then did task-centered practice gain the status of an established practice model, and one that could believably be credited with paradigmatic changes? Six reasons can be identified. First, it could plausibly claim to have both continuity with previous valued practice positions while offering a significant break—in Polanyi's terms, to trust the tradition while at the same time claiming an independent position from which this tradition may be reinterpreted. Fortune observes that "most practitioners could find some aspect of the task-centered model that was similar to what they were already doing, and this familiarity tempered some of the perceptions of how radical the model was" (2012, 24). She describes it as having the hallmark of being an "intermediate approach" (31) that enabled practitioners to change without having to sign on to something that seemed an abandonment of existing valued positions.

Second, and by way of extension of the previous point, it fitted into wider influential narratives without seeming a mere subcategory of something bigger. Fortune (2014), in her role as guardian of the tradition, argues that the model belongs to the family of evidence-driven and linear intervention models while being distinctively American and not the British import of evidence-based practice.[13] "It has consistently modelled an empirical orientation to practice" (Rooney 2010, 198). Rooney makes an interesting remark in this context regarding how task-centered practice has proved itself as a "metamodel for how social work models or approaches could be constructed" (197–198).

Task-centered practice gains strength, third, by virtue of being able to bolt on otherwise unconnected discourses. In the late 1960s and early 1970s, for example, there were emerging concerns that social work practice almost routinely disregarded or explained away the views of the client (e.g., Mayer and Timms 1970; cf. Shaw 1976). While there is little evidence to align Reid with broader movements of this nature, it was significant that task-centered practice could be promoted as "the first social work practice model to focus on the *client's* actions" (Videka and Blackburn 2010, 188). Also, problems tend to be taken at face value, such that "additional meaning is not read into the client's statements and target problems are not reinterpreted to mean other problems" (189). The dimension of reciprocal work tasks for social worker and client also implied a practice value enabling commentators to say it "has embodied the social work value of self-determination and supported client empowerment and facilitating strengths" (Rooney 2010, 198) and "was client centred and client driven even before empowerment and strengths-based practice became the mantra of the profession" (Videka and Blackburn 2010, 191).

Fourth, the surrounding rhetoric, largely put forward by advocates other than Reid himself, succeeded implicitly in aligning task-centered practice to the recurring Enlightenment periodization of social work, with its accompanying images of social work as progressively more rational, systematic, and susceptible to measurement and evaluation. The unexamined metaphorical language also serves to convey the image of social work as a matter of substance and solidity—it is a "model" that is evidence *based*, research *based*, scientifically *based*, and so on. It has an architectural foundation. It is not something added to the top or the exterior by way of decoration but something fundamental.

Fifth, and perhaps least likely to be noticed, it reinforces a narrative of U.S. intellectual and scientific hegemony. Other countries appear in fleeting glimpses and always as beneficiaries of the U.S. invention—a view generally more or less accepted by writers in other countries. The most visible way to observe this is through a scan of the citations in key papers. In a key article by Fortune (2012) there are 116 cited sources, of which 105 are, to my knowledge, U.S. writers. The account is almost entirely a U.S. history. Sackett's evidence-based medicine from Britain is the only identified influence on the United States, and it is one that Fortune distinguishes from the empirical clinical practice movement, to which she aligns task-centered practice. A similar picture can be seen in the other articles referred to above, but given that this characterizes almost all U.S. social work writing, it is not a unique feature on invention scholarship.

Finally, task-centered practice can be readily converted to instrumental prescriptions. In general terms it has a "method," in the sense (not pejoratively) of cookery—"it had a technology" (Fortune 2012, 30). Associated with this it was susceptible to development and thus to regular modifications and partial relaunches. Reid was probably a gradualist, and the opportunity for incremental development of the model may have suited his temperament. Rooney observes how it has "proven quite flexible" (2010, 198). He remarks on the paradox that "one of the strengths of the task-centered approach has been its acknowledgement of limitations," for example, as not being an all-purpose model (198). Links to Reid's doctoral program gave it developmental capacity (Fortune 2012, 27).

Interests and Controversies

The previous account of ways that task-centered practice may be seen as an invention bridges the major part of the chapter with the following pages. Having begun by considering different ways in which social work science has community characteristics, the understanding of inventions introduces ways that elements of difference and even disagreement may occur. We will explore this issue by examining two related questions: How do social interests shape science? And in what sense is it meaningful to speak of social work as marked by scientific controversies?

Interests

"Interest" is something of a catch-all term—having a right to something, having a stake in the outcome of something, a feeling of concern regarding something, and a chosen or preferred preference for something. The word is used in this book mainly to deal with two questions. First, how convincingly may we, by doing science, claim to understand something or someone's interests as well as or better than do they? Second, what should we make of the argument that the processes, results, outcomes, and applications of science are shaped as much, or even more, by collective interests than by rationality? The first question is explored in chapters 7 and 8; the second question is briefly considered in the following paragraphs. It provides a prelude to talking about social work science controversies.

We came across the dispute between Yule and Pearson regarding the measurement of association between nominal variables in chapter 3, where we saw in passing the argument that to grasp what was at issue we need to take into account the cognitive and social interests of the two parties. To express this in ways that connect with things said regarding task-centered practice, we are asking if statistical concepts are discovered or invented.

At one level it is a commonplace to insist that science is a social enterprise and that there are social causes, shaped by interests, of what scientists believe. A stronger version of this position would concede that "the content of 'good' science as well as 'bad' can potentially be affected by its social context" such that "the growth of scientific knowledge cannot be understood entirely in terms of 'its own laws'" (MacKenzie 1981, 3). At one level this is still a relatively modest position, to the effect that the production of new ideas is socially influenced, but these ideas are then judged by scientific criteria—a position not unlike that taken by Weber. However, it is possible to radicalize the argument and reason that not only the production of new ideas but also the process by which those ideas are accepted or rejected can be affected by social factors. MacKenzie sees no reason to limit influences to those within the scientific community but includes society at large.

The dividing line is whether one includes the beliefs of scientists about the content of their science as causes. As Phillips (2000) expresses it, the issue is whether it helps distinguish the politics of science and the truth of science, and if knowledge claims regarded as true should be explained differently from knowledge claims regarded as false. Witkin seems to say no to both questions when he suggests that "there is no intrinsic reason, apart from the interests of particular groups . . . to limit knowledge claims to certain criteria" (1999, 7), thus apparently adopting the "strong"[14] position that we should seek to explain "true" positions in just the same way as we should seek to explain "false" positions. To the strong constructivist, the distinction between internal and external reasons/explanations of belief has no meaning because the rational is always social. This way of thinking has some similarities to Foucault's notion of *episteme* and Kuhn's early position on paradigms.

To lay my cards on the table, I accept the distinction between internal and external causes but concur with Phillips when he says that it "is a rough or permeable distinction rather than a watertight one" (2000, 206)—although I may give greater weight to external reasons/explanations.

Back to the Pearson versus Yule controversy, MacKenzie's argument is that Pearson always "sought a coefficient of association directly comparable with the correlation coefficient of interval variables" (MacKenzie 1981, 159). But for Yule nominal categories are "naturally discrete classes," and to apply a coefficient that has as its basis an assumption of underlying continuous variables was absurd. Their positions were incommensurable because they were based on different goals. For Yule it was a question of "simply trying to summarise the degree of dependence manifest in the given nominal data" (167). He was happy for there to be different ways of doing this, which to Pearson made no sense. For Yule different values would be true depending on where the discrete boundary lay e.g. between tall and short people, but "for Pearson this invalidated Q" (168).

Pearson was working within his eugenic position of heredity, and he was "constructing a mathematical theory of descent" (168). He wanted to be able to say what the coefficient of heredity was, and that could only be done by assuming continuous variables. Yule had no commitment to eugenics—his attitudes from correspondence seemed to be "a mixture of indifference and hostility" (173). "The eugenic concerns embodied in Pearson's work led to a sophisticated and elaborate theory constructed round a specific goal, while in Yule's work more diffuse concerns led to a looser approach that embodied goals of a more general nature" (175). Later positions were more pluralist and thus closer to Yule. Yule's test has been favored by sociologists. But Pearson's construction of models to fit data has survived and developed. Yule was opposed to Pearson's positivism and also of the cult of measurement. He was also "wary of too close a connection between 'pure' science and its applications" (181).

Historical analysis of this kind points to seeing sciences "as oriented to secular and definable goals, and not to see them merely abstractly as the pursuit of truth" (216) and of science as reflecting a goal to expand human capacity to predict and control the world. "Successful statistical procedures enhance the potential for control" (217). "Knowledge must be analysed as a resource for practice, and knowing must be seen as a process" (225).

While this kind of argument is deeply important for how science is understood and deployed in social work, it needs to be taken with considerable care. MacKenzie admits that there is an interpretive problem. How can we relate society and its structure to the beliefs of particular people? We may study empirically, looking for regularities in the relationship, but such evidence is of little value because the relationship will be tentative. Also MacKenzie is not arguing that the views of any one person were socially determined. "Interests" are positioned as a way of explaining knowledge generation—not in a determinist way but as influencing knowledge production. As Woolgar (1981) expresses it, scientists can be seen as "constantly engaged in monitoring, evaluating, attributing (in short, in 'accounting for') the potential presence or absence of interests in the work and activities both of others and of themselves." However, "we have as yet little appreciation of the way this kind of work is done by scientists" (371), and he argues that "far greater reflective attention be given to the *explanatory form* of interests explanations" (373). Woolgar extends his reservations to the way "interests" connect to arguments about the effect that "context" has on scientific work and suggests "here again is a notion mobilized by historians and sociologists with little regard for its currency as resource which is constitutive of scientists' argument; the concept is employed as unproblematically fixed (rather than constructed)" (374; cf. Shaw 2010).

Controversies and Social Work

An example of the challenges posed by developing an adequate explanation of interests can be seen in chapter 3 (example 3.2, on social work research methods and practices). While it seems relatively clear that interests (presented there as discourses) have contributed to social work science much more centrally than is usually appreciated, a major deconstructive force of that example is to question any homogenous or unified notion of interests. Indeed, the apparent discontinuities within the different discursive arguments for good social work research provide a bridge to understanding the place of controversies in social work science.

But is social work marked by controversies? The question may seem naïve, but it reminds us again of the need to be clear about just what is being said. Also, once one looks for discussion of the question it appears that the nature of controversies is too little considered in social work. This is attributable in part to how the rationalist tradition, by its assumptions, sees controversies as abnormal, deviant phenomena. Thus, critique is central but is seen as how the scientific community should disclose mistakes and eliminate disagreement. "Thus scientific disputes are reconstructed to become amenable to a rationalist image of scientific development," which in turn serves to set up "a *cordon sanitaire* between science and any other forms of knowledge" (Brante and Elzinga 1990, 34). The position taken here is that the occurrence of controversies is not a question of error but "an indicator of . . . deep-rooted tensions which go to the very heart of science, both with regard to its character and function" (35).

Taking arguments about the social dimensions of science as an example, Latour (1992) presents controversies as a tug of war. On one side we have radicals and progressists, on the other reactionaries and conservatives.

1. Radical: scientific knowledge is constructed entirely out of social relations
2. Progressist: science is partly constructed out of social relations, but nature somehow "leaks in" at the end
3. Conservatives: although science escapes from society, there are still factors from society that "leak in" and influence its development
4. Reactionaries: science becomes scientific only when it finally sheds any trace of social construction

Returning to Kuhn, it is possible to categorize controversies as of three kinds. First, there are preparadigmatic controversies. He prefaces *The Structure of Scientific Revolutions* by remarking that when he encountered social scientists he was struck, in comparison with natural scientists, "by the number and extent of overt disagreements between social scientists about the nature of legitimate scientific problems and methods" (Kuhn 1970, viii). Second, there are anomalies within a given paradigm. These in turn are of two kinds—those that can be treated as puzzles (albeit often very demanding ones) that can be tackled within the assumptions of the paradigm and those more fundamental problems that cannot be resolved and that accumulate toward a crisis.

Brante and Elzinga (1990) define a controversy as "primarily concerned with contending knowledge claims, where at least one of the parties has a scientific status" (36). "Controversy creates interaction; thus it signifies unifying as well as divergent tendencies between groups of antagonists." A controversy also "has a certain endurance in time and space. . . . In general (although not always) a controversy exists over a longer period of time and divides groups of people" (36) rather than being just between single individuals.

Collins, Godin, and Gingras (2002) suggest that scientific controversies have three stages:

1. A new fact is claimed as result of scientific work.
2. Because of the interpretive flexibility of the data, and because replication is never simple, debates and conflicting views, supported by other different experiments, emerge.
3. Debate may be long lived or short lived but eventually ends with closure.

"Interaction between scientists is the key to understanding controversies. . . . Scientific knowledge is thus a set of arguments (experimental or theoretical) that have *survived* objections—with no absolute certainty that they will resist *future* attacks. . . . A scientific debate terminates when other participants can no longer argue" (145–146, 148).

This is not a comprehensive way of thinking about the history of controversies. Different types of controversies occur because (in ways that echo Kuhn) there are different kinds of conflicts. These may be with regard to facts (to do with what is observed), theory (when two or more theories claim to explain the same phenomenon), or principle. "Controversies of principle often involve entire perspectives and basic components of world views" (37). There are occasions where scientific, moral, and political principles interact—a controversy of both knowledge and action, including a dispute about what should be done.

The relative importance of internal and external validity in evaluation research affords an example of a controversy that, contrary to Brante and Elzinga, did focus for a time primarily on two figures—Donald Campbell and Lee Cronbach. The reason for including this example is partly because the issues Cronbach and his colleagues raised have been little noticed in social work and partly to illustrate how decisions about seemingly "technical" issues of measurement are fundamentally shaped by wider value positions. Hence I have not set this as a "balanced" statement of each "side."

EXAMPLE 5.2
Lee Cronbach on External Validity

"We emphasise not form of inquiry but relevance of information" because "under many circumstances, the emphasis on assessment of outcomes of a supposedly fixed program runs counter to the aims of understanding the problem and rendering better service" (Cronbach et al. 1980, 216). Taking research and evaluation as accountability "is both a limited view of the reasons for program success or failure and a limiting view of how evaluation can best be used to bring about improvement. Evaluation is not best used, we think, to bring pressure on public servants" (17).

Cronbach and his colleagues are critical of injunctions to evaluate against clear and measureable goals, with assignment, and reliable and objective measures that sustain internal validity. "We do not consider it reasonable to separate the effects of the program from the rest of the client's experience." Speaking of schools evaluation, they say a "program that appears superior to a rival program in isolation may be inferior when each program is embedded in the regular sequence of school experience" (217). For example, "after the experimenter with his artificial constraint leaves the scene, the operating program is sure to be adapted to local conditions" (217).

Hence, "external validity"—validity of inferences that go beyond the data—is the crux of social action, not "internal validity" (231). Internal validity gains may be feasible "to some extent, but relevance is likely to suffer" (231). While always favoring relevance over precision, they do not see them as choices.

An internally valid study is one in which the "statistician's conclusion"—the inference from sample to population—is beyond challenge. That ideal is most surely achieved by a comparative experiment with controlled assignment. Internal validity can claim priority when the investigator addresses a summative and causal question, not otherwise. This is the position of all Campbell's writings on the subject, but it becomes explicit only at the end of the Cook and Campbell monograph (1979, 343): "Though random assignment is germane to the research goal of assessing whether the treatment caused any observed effects, it is conceptually irrelevant to all other research goals." Internal validity, however, is not of salient importance in an evaluation. What counts in evaluation is external validity, that is, the plausibility of conclusions about one or another (Cronbach et al. 1980, 314).

Part of the problem is that "ideas face an up-or-out decision much too early." He insists "progress requires that we respect poorly formed and even 'untestable' ideas." In a nice phrase he says, "we should be stern only where it would cost us much to be wrong" (Cronbach 1986, 86). "We shall not advise the evaluator to avoid opportunities to be helpful; he is hired to improve public services, not to referee a basketball game" (Cronbach et al. 1980, 18).

A remaining difficulty is that no amount of categorizing of kinds of controversies will save us from metacontroversies of disagreement about what sort of controversy is at stake in any given instance. Those of a certain age may recall the heat, cut, and thrust of the debate following the publication of *Girls at Vocational High* captured well in Mary Macdonald's response essay, to which we return in chapter 6 (Meyer, Borgatta, and Jones 1965; Macdonald 1966). Even accepting the distinction between controversies of fact, theory, and principle, it is not possible clearly to separate different types, in that facts are theory and/or principle dependent, so it is more a difference of degree and focus. "Theories and conceptual frameworks tend to generate 'their own' facts" (Brante and Elzinga 1990, 43). There remains the issue—itself the focus of controversy—as to whether controversies are socially or cognitively generated, reflecting the difference between rationalist and relativist approaches. "If a controversy can be analysed as 'pure,' that is, generated by scientific arguments only . . . then there seems to be some basis for maintaining the thesis of the autonomy of science." But "there is always a social aspect even in the most internalist of controversies" (38–39).

Agreeing About Disagreements

In closing this chapter we should say something briefly regarding how disagreements ought to be handled. Three kinds of work are needed—empirical, cognitive, and social.

Empirical

Kirk and Reid (2002) correctly stress the need for treating questions of science empirically. This applies to understanding epistemologies adopted by social workers as well as acknowledging that "the bottom line for research utilization is what actually happens in the field among practitioners" (194). Brante and Elzinga suggest that controversies may be studied epistemologically, descriptively, or in terms of their political context. Empirically we may seek to grasp their emergence, including how contending claims for jurisdiction over a specific field or problem arise; who is regarded as having the right to speak; who is the real expert. The development phase of a controversy points to understanding alliance seeking and how resources are mobilized. In studying their termination, they may be resolved by scientific argument (cf. Oakley 1999), something nonepistemic may close the debate, it may be abandoned, or protagonists may go separate ways, for example, by specialization.

Cognitive

Cognitively, I would advocate three stances. First, the value of a falsifying perspective. Phillips knows there have been philosophical rejoinders to Popper, though he thinks that the position can have practical gains without being philosophically complete. Even if Popper is wrong, following him is benign, and the errors made would not be serious. "It will incline people to accept criticism, state their claims with clarity, cast their research design in fruitful ways, open their theories and hypotheses to empirical test" (Phillips 2000, 141). He laments, in terms that have a measure of force, that "the tendency to search for 'confirming' rather than disconfirming or refuting evidence is particularly strong in research that uses qualitative methods" (144) and is often marked by a "mental set" that directs the observations. He suggests a Popperian recasting of Type I and II errors as:

Type I: tentative maintenance of a false hypothesis
Type II: tentative rejection of a true hypothesis

He suggests that Popper might say "we can never completely insulate ourselves against either of these types of errors; *all* our knowledge is fallible, and much of what we think to be true is, in fact, likely to be fallible" (153).

Second, we should retain, despite the manifold difficulties it brings, the distinction between rational and social causes in science, between epistemic and nonepistemic criteria for good social work science and research. I have implied this position earlier in the book and developed it at greater length in earlier work (Shaw and Norton 2007). In terms of assessing quality, the social work community should not drill down to very specific instrumental standards. Also, a standard that is "fit for purpose" in social work science should include guidance on how different stakeholder communities should apply quality judgments. This is not the same as saying that a practitioner/policy/service-user version will be "thinner" or involve less expertise than a scheme tailored for universities. Nor does this justify a free-for-all relativism. All stakeholders should sign up to the broad dimensions and standards, but their application should always leave scope for flexibility and local relevance. Quality should be based on justification of both inner and outer science considerations taking into account the context (including the aims and objectives) in which the research is taking place.

Third, there are real controversies in social work science, ones that can never be restricted to rational, cognitive issues. It may be feasible to find a *via media* that incorporates important aspects of, for example, Kuhn and Popper by regarding science as consisting of competing fragments of Kuhnian "normal science" or "research programs." Instead of seeing research programs of this kind as running in series, one after another, as the paradigm argument requires, research programs should be thought of as running in parallel and in competition.

Social

Practitioners of social work science should avoid sentimentality of the kind criticized in chapter 3 and adopt a moderate skepticism, especially about their own work. The philosopher most associated with skepticism is perhaps René Descartes. To provide a firm ground for knowledge, he began by "doubting all that could be doubted" (Grau 2007, 195), for example, by asking what if he were only dreaming. Contemporary films, such as *The Matrix,* explore this idea via speculations about technology. Hume saw no way out of Descartes' problem and thought that we have to live with it, on the grounds that we can be philosophical skeptics yet sustain a practical belief in reason. Does (moderate) skepticism lead to paralyzed inaction? Not necessarily. It is possible to be philosophically skeptical yet practically committed to action.

I am uncomfortable with how disagreements often are conducted in social work. There is an overly frequent tendency to resort to what Raymond Williams called "swearing." We commence in attack mode. In addition, we too rarely source arguments in the writings of those who are proponents of that with which we disagree, and too often we take as adequate the work of those with whom we are sympathetic. There is also a tendency to voice positions via a reductionist analysis of those with whom we disagree. Whether I have followed my own prescriptions is a matter for others to judge. I have suggested ways that I find quantitative, largely intervention-based research as fruitful, and I have debated with those whose stance on balance I find congenial (Shaw 2012b, 2014e). In so doing I assume that I can—and probably must—talk with (and not at) those with whom I profoundly disagree, on the assumption that in principle we can hear each other. And that I should respect and assume that their positions are held with integrity and have as much right to be heard as mine. In accord with this, I have long taken the rather challenging (for me) approach of asking people with whom I deeply disagree to comment critically on my draft papers. I spend much time thinking about what I am walling in or walling out. I have remarked somewhere that, as with the poet Robert Frost in "Mending Wall," I have no time for hunters who leave "not one stone on a stone . . . to please the yelping dogs." I do aspire to provoke conversation and debate. I doubt it would make consistent sense to do any of this if I believed that those with whom I disagreed lived in another paradigmatic world.

Taking It Further

Reading

Brante, T., and A. Elzinga. 1990. "Towards a Theory of Scientific Controversies." *Science Studies* 2: 33–46.

Dearing, J. W. 2009. "Applying Diffusion of Innovation Theory to Intervention Development." *Research on Social Work Practice* 19(5): 503–518.

Kuhn, T. S. 1970. *The Structure of Scientific Revolutions*. Chicago: University of Chicago Press.

I do not know of any single source that covers the general ground in this chapter. But Dearing's article is an excellent stimulus for thinking about the spread of knowledge. Thomas Brante is one of the few who have attempted to theorize scientific controversies. Finally, suspecting that Kuhn is more named than read, we cannot go further in this book without recommending a cover-to-cover reading of his classic text. The various later writings about Kuhn scattered throughout this book then will yield more fruitful reading. But this chapter stands as perhaps the only overview, thus far, of the social organization and practices of social work science.

Task

Identify a position in relation to social work science on which you hold strong and considered views. Locate two high-level articles—one that takes a similar position to yourself and one with which you have disagreed. Consider the strengths of the latter and the limitations of the former.

[6]

Social Work Science and Evidence

Chapters 6 to 8 take up three central themes in the nature and purpose of social work science: science as evidence, science as understanding, and science as justice. Beginning chapter 6 with the view that social work science offers a form of "foundationalism," I will then suggest how science serves to define the nature of social work as professional work, thus providing a foundation for viewing evidence-based practice as located close to the core of social work. I then sketch the meaning of scientific or evidence-based practice and discuss how arguments are made in their support. I will connect that to how ideas of accountability are central to this view of social work and science.

The chapter then moves on to see in more detail ways in which commitments to evidence as informing practice have developed and responded to counteracting positions. I will touch in passing on how realist views of science offer important modifications of evidence as science. Feminist positions frequently have been seen as running directly counter to evidence-based practice. We will see that this is an overly simple understanding.

I then reflect on the relationship between commitments to evidence and the methods of science. The chapter closes with a general assessment of evidence as the central foundation of social work science.

Questions of evidence and science take us close to the heart of how the purpose of social work is understood in North America, large parts of Europe, and, indeed, in almost all countries where social work has accomplished some degree of professional status. In exploring the issue of how the idea of and commitment to notions of best evidence shape social work science, I stress that this chapter is not a review of outcomes or intervention research and their associated methodologies. Nor am I dealing in a comprehensive way with the evidence-based-practice movement or its close relatives, empirical practice, scientific practice, and the like[1]—though they inevitably have surfaced elsewhere in the book through, for example, discussions of science "inventions" in social work. To put my cards on the table, however, my general view of the empirical practice movement and its cognates is that "the gains achieved have been partial, uneven, and not always clear" (Reid 1994, 180). By way of preamble, I briefly consider the view that social work science offers a form of foundationalism and its consequences for how science evidence is regarded.

Foundationalism and Evidence

Leplin wryly remarks, "like the Equal Rights Movement, scientific realism is a majority position whose advocates are so divided as to appear a minority" (1984, 1). Realism in science is usually assumed to demand an absolutist view of the role of scientific inquiry, referred to as foundationalism.[2] This involves some form of a correspondence view of truth, that is, that science can represent realities independent of the inquiry process. In its strong, "naïve" form, realism retains the positivist adherence to the role of observation as the discloser of the true nature of reality and a confidence in the objectivity of the researcher. However, the use of the term "positivist" as exhausting the scope of foundationalism is misleading, and its elasticity renders it of limited value.

> In its most extreme form, foundationalism presents research, when it is properly executed, as producing conclusions whose validity follows automatically from the "givenness" of the data on which they are based. This may be assumed to be achieved by the "immediacy" of the conclusions or by methodological procedures that transmit validity from premises to conclusions.
>
> (Hammersley and Gomm 1997, 2.3)

Foundationalist science gives a central role to the countering of bias. Traditional understandings of bias depend closely on notions of truth, validity, and objectivity. There are two senses to bias. First, it "commonly refers to systematic error: deviation from a true score, the latter referring to the valid measurement of some phenomenon or to accurate estimation of a population parameter" (Hammersley and Gomm 1997, abstract). It may also have a more specific sense of a particular source of systematic error: "A tendency on the part of researchers to collect data and/or to interpret and present them in such a way as to favour false results that are in line with their pre-judgements and political or practical commitments" (1.7).

Hammersley and Gomm find both uses problematic because they seem to rely on foundationalist epistemological assumptions, though for Hammersley "minimising bias is, for me, the core of scientific inquiry" (Hammersley 2005, 93). They opt for a notion of collegial accountability. Strong versions of realism are present in some varieties of qualitative research. For instance, Blumer argued that a constructivist view

> does not shift "reality" . . . from the empirical world to the realm of imagery and conception . . . [The] empirical world can "talk back" to our picture of it or assertions about it—talk back in the sense of challenging and resisting, or not bending to, our images or conceptions of it. This resistance gives the empirical world an obdurate character that is the mark of reality.
>
> (Blumer 1969, 22)

Understanding foundationalism also impinges on questions of the relationship of theoretical and empirical bases of social work science and of discussions of falsificationism. A difficulty with strong foundationalist positions follows from the realization, set out in chapter 1, that the relationship between theory and observation has been shown to be more complex than previously thought. It became clear that theories are "underdetermined" by nature, such that we are never able to say that we have the best theory; a variety of theories can be constructed that are equally compatible with the available evidence. This has led to the rejection of naïve strong forms of foundationalism, with their implication that research findings of indisputable validity can be a foundation for policy.

One commonsense understanding of falsification might proceed by saying that *theories* are always fallible but that the *empirical base* for theories is not fallible. Therefore falsification entails testing fallible theories with nonfallible facts. Lakatos (1970) labels this position "dogmatic falsificationism." He rightly rejects the caricature that "science grows by repeated overthrow of theories with the help of hard facts" (97). Neither was this Popper's position. As an antideterminist he emphasized, even celebrated, the uncertainty of knowledge. Knowledge for Popper never equaled proven knowledge because for him knowledge is not provable. When Popper spoke of an empirical base, experimenting, observing, applying theories, and eliminating or falsifying theories, there were always at least invisible inverted commas around each word, indicating that they are fallible. Popper advocated all of these activities but did so aware of the approximations to knowledge and of the fallibility and risk at every step in our conclusions. Exhortations to social workers to "falsify" ought to assume the same inverted commas.

A vital conclusion follows. "If a theory is falsified, it is proven false; if it is 'falsified' it may still be true" (Lakatos 1970, 108). "Eliminated" theories—and research findings—may not be false. Social work science entails "falsifying," not falsifying. This is very far from naïve, dogmatic objectivism. Indeed, this may tempt us to drop out in the belief that testing our conclusions is meaningless. But Popper is severe on this point. He insists in a memorable phrase that our hypothesis "must be made to stick its neck out." But we must live with the uncertainty of what we know. We may reject an explanation, but we cannot disprove it. The real debate is not whether social work is scientific but whether it is "scientific."

Moderate positions have been expressed within social work. For example, Kirk and Reid (2002) rescue these debates from short-termism by retaining a historical rootedness. They also permeate their discussion with strong awareness of the literature on the philosophy of knowledge, methodology, and some key evaluation theorists. In addition, their reserve about easy solutions yields an even-handedness to their discussion of the criticisms leveled against scientific practice. They are ready to acknowledge that theoretical assumptions *are* implicit in scientific practice. In a helpful mediating position, they say, "it may make sense to construe scientific practice as a 'perspective' on intervention," and they talk about "being willing to accredit client ideas about measurement, data collection and the like that might not fit conventional research notions" (89). They even suggest that a scientific practice perspective "could be used with advocacy research, even though the practitioner-researcher might need to forgo his or her 'neutrality' " (89; cf. Reid and Zettergren 1999).

Madge cites C. Wright Mills to suggest that "the accumulation of knowledge may not necessarily be the primary function of science" and that hence we are "free to regard scientific activity as containing its own rewards, with knowledge as a residue rather than as an end" (Madge 1953, 2), and suggests that "to postulate an objective social science is to ask for something which is probably unattainable and may even be undesirable" (6) and even that "in no science is the pursuit of objective knowledge more futile than in social science" (2). "In a crude way we are sometimes taught to conceive of science as comprising three almost mechanical stages, namely observation-hypothesis-verification. This is almost wholly misleading." Disciplines develop aids to improve observation, but "no such improvements can be won without some corresponding loss. In almost every case the loss is a certain restriction on the freedom of observation" (11). In metaphorical terms he remarks, "as the magnification of a telescope is increased, so the field of vision is restricted" (11).[3] In short, then, "while the abandonment of foundationalism requires us to recognise that research will inevitably be affected by the personal and social characteristics of the researcher, and that this can be of positive value as well as a source of systematic error, it does not require us to give up the guiding principle of objectivity" (Hammersley and Gomm 1997, 4.12). We can maintain belief in the existence of phenomena independent of our claims about them, and in their knowability, "without assuming that we can know with certainty whether our knowledge of them is valid or invalid" (Hammersley 1992, 50).

Professions and Science

The status and authority of "evidence" in social work science is intertwined with how social work is viewed as a profession and its relation to other professional fields. To take a view regarding professions is premised on assumptions regarding social and individual problems. Abbott rejects what he regards as "two easy, and erroneous, views" of the relation of objective facts regarding problems and subjective perceptions. To those who take the stance that problems, such as those dealt with by social work, are objective facts and that varying perceptions are either errors or ideologies, or that problems have no objective reality but are simply constructions, he responds: "neither view works" (Abbott 1988, 37). Constructions of a problem—for example, mental health—are always "bounded by the objective properties" (38). This is how he develops his view of professions as having several types of objective foundations—technological, organizational, and, though they "should not be overestimated," "natural objects and facts (39)—though ones that are always vulnerable to changes. They also have subjective qualities. These too are vulnerable—for example, to the activities of other professions (in the case of social work this is, more than any others, to changes in medicine, though in parts of Europe it includes social pedagogy).

Abbott spent time reading the history of professions as "a history of turf wars" (1995, 552), seeing the boundaries of social work as shaped by conflict over boundaries. But he later suggests that this does not explain quite why and from where social work came. He suggests that "social entities come into existence when social agents tie social boundaries together in certain ways. The first things are the boundaries. The second are the entities" (555). Local boundaries and groupings always come first, before the wider entity.

> The standard account of the formation of social work views the profession as arising out of a turf competition between the charity organization societies with their "scientific" ethos of casework, and the settlements with their chaotically comprehensive services and their broad social agenda. The view is that the settlements more or less lost out to the new scientism of Mary Richmond and others. The settlements' broad interests in reform and preventative services were replaced by the narrow, vocational, casework-centered approach of the social work schools.
>
> (556)

Over against this he argues that in the various subareas, such as probation and family work, "boundaries began to emerge between different kinds of people doing the same kinds of work, or between different styles of work with roughly similar clients, or between one kind of workplace and another. . . . They were not boundaries *of* anything but, rather, simple locations of difference" (557). "Thus, I come to the notion that social entities actually emerge from boundaries" (558). Child abuse is one obvious example of vulnerability to—yet influence upon—the activities of others (example 6.1 links text quoted previously in the book with other elements of Hacking's argument).

Abbott (1995) develops the argument that the subjective qualities of a professional task "arise in the current construction of the problem by the profession currently 'holding the jurisdiction' of that task." His position in summary is as follows: "In their cultural aspect, the jurisdictional claims that create these subjective qualities have three parts: claims to classify a

EXAMPLE 6.1

Child Abuse and the Professional Culture of Social Work

Child abuse not a constant. Ian Hacking, the Canadian philosopher of science, observed in 1991 how the abusers' "own sense of what they are doing, how they do it, and even what they do is just not the same now as it was thirty years ago" (254), such that "no one had any glimmering, in 1960, of what was going to count as child abuse in 1990" (257). "Since 1962 the class of acts falling under 'child abuse' has changed every few years, so that people who have not kept up to date are astonished to be told that the present primary connotation of child abuse is sexual abuse" (259).

Hacking elaborates the ways that child abuse connects to other "vexing issues." Morality has become part of the center of how crimes are rated. "Our whole value system has been affected by the trajectory of child abuse in the past thirty years" (259). Furthermore, "without feminism there is little likelihood that the idea of child abuse would so quickly have absorbed the notion of sexual abuse of children. Wife assault and child assault have become assimilated and the entire phenomenon of child abuse seen as one more aspect of patriarchal domination" (260). Then again, "had not child abuse finally become a focus in the 1960s of vast public concern, the question of children's rights would be almost unknown" (260). The role of the courts is also central through, for example, fundamental shifts in how the testimony of children is both heard and treated. The medical connection is also illustrated in ways that "views on the causes and prevention of child abuse have determined, to a great extent, the class of events that are labelled abuse" (261).

Source: Hacking (1991).

problem, to reason about it, and to take action on it: in more formal terms, to diagnose, to infer, and to treat. Theoretically, these are the three acts of professional practice" (40). He elaborates that

> professionals often run them together. They may begin with treatment rather than diagnosis; they may indeed diagnose by treating as doctors often do. The three are modalities of action rather than acts per se. But the sequence of diagnosis, inference and treatment embodies the essential cultural logic of professional practice. It is within this logic that tasks receive the subjective qualities that are the cognitive structure of a jurisdictional claim.
>
> (40)

Central to the work of professions is how "science, with the broader, related phenomenon of formal rationality, has become the fundamental ground for the legitimacy of professional techniques. In the value scheme on which modern professions draw, science stands for logic and rigor in diagnosis, as well as a certain caution and conservatism in professional therapeutics" (189).

While science had its main effect in transforming professional social structure, for example through universities, "it changed legitimation of cultural structure as well," through commitment to rational efficiency (190), though there has often been tension between the value of general learning as against specialist learning and also between differing views on the importance of character as against learning. But the "major shift in legitimation in the professions has thus been a shift from a reliance on social origins and character values to a reliance on scientization or rationalization of technique and on efficiency of service" (195).

Abbott is one of a number who have come to reject the listing of traits as defining professions. In a radical version of such positions, Bourdieu's concept of "field" replaces that of profession. Brante, however, does not want to abandon the notion of "profession." He picks up one trait—"knowledge"—and proceeds to work on what is entailed in "professional knowledge." He asks, "What distinguishes professions from other occupations, simultaneously denoting what professions have in common?" and "What *type* of knowledge generates the skills and practices we call professional?" (2011, 5, 9). The synopsis of his argument is that professions are the "inter-mediators and appliers of the highest knowledge within specific social domains. There is no higher authority, no deeper source of knowledge and action to turn to" (9). "Currently, the highest modern knowledge is produced and distributed at the universities and is called scientific. Hence, contemporary professions are science-based occupations; their practice involves applied science." In the context of the proliferation of universities and programs "we attempt to delineate a certain kind of scientific knowledge as the basis for professional practice" (10). My reading of Brante is that he is saying something regarding the identity of professions rather than arguing that in any evidential way they base the core of their work on the results of "science."

A core element of his position is to see what he calls "ontological models" as acting as bridges between science and profession. They "constitute bridges between theory and observation within a scientific tradition; models coordinate theory and data" (11) in ways that remind us of Kuhn's paradigm and Hacking's "styles of reasoning." "Together, these functions provide the presuppositions for an epistemological break with everyday notions of reality" (13). On this basis he argues that "*scientific research and professional practice are governed by a shared basic model that breaks with everyday knowledge, with 'common sense'*" (14). Professions are "*occupations conducting interventions derived from scientific knowledge of mechanisms, structures, and contexts*" (17). "A profession obtains its status from a central base, that it is a truth regime. Because of its scientific base, a profession is the ultimate link to 'truth'; there is no higher authority. This and only this is what makes professions unique" (19).

Evidence-Based Practice

Western thought tends to see history in a linear form. This is related to fundamental ideas of how we see cause and effect. There are varieties of such thinking around today, and in some unexpected places. To offer two strikingly different instances, I suggest that Denzin's successive arguments about "moments" in the history of qualitative research have hints of a view of positive knowledge progression (cf. Denzin and Lincoln 2005).[4] By contrast, a history of evidence-based practice unashamedly advanced the case that "the first shift in social work, from a morally based to an authority-based practice paradigm, was finalized by the early 1930s. The second shift, which is an attempt to transition from an authority-based to an empirically based practice paradigm, was initiated in the late 1960s and 1970s but has not been completed" (Okpych and Yu 2014, 5).

Morally, authority based, empirically—these terms are almost identical to Comte's theological, metaphysical, and positive phases. The brilliant journalist Simon Jenkins[5] captured this attitude perfectly in his remark that "some people think that living in the sixteenth century is no excuse for not having read *The Guardian*."[6]

Sackett's definition of evidence-based practice as "the conscientious, explicit and judicious use of current best evidence in making decisions about the care of individual patients" (Sackett et al. 1996, 71) is cited often enough in human services and social work literature to give the impression that discussion about how to define evidence-based practice is redundant. This would be unhelpful. We should distinguish between evidence-based practice as a model of intervention that has been demonstrated effective and evidence-based practice as a way of practicing, using critical thinking. The first of these sometimes has been divided into two to distinguish basing practice decisions on empirically based evidence as to the intervention strategies that are likely to produce the desired outcomes and evaluating the implementation of these interventions to ensure they are being implemented as intended (e.g., Roberts and Yeager 2004, 5). The second characteristic, critical thinking, focuses more on professional judgment. It "denotes an approach to evidence, argument and decision-making which is closely related to scientific reasoning" (Macdonald 2000a, 80) and has been said to include:

> 1. Assessing accurately the quality of evidence
> 2. Recognizing and countering common fallacies of reasoning
> 3. Recognizing how affective and cognitive biases can adversely influence professional judgment
> 4. Establishing "a *modus operandi* which promotes accuracy or truth over winning the argument"
>
> (Macdonald 2000a, 80)

Macdonald brings these various aspects together when she says that evidence-based practice represents "an approach to decision-making which is transparent, accountable, and based on a consideration of current best evidence about the effects of particular interventions on the welfare of individuals, groups and communities" (2000b, 123).

The characteristics of evidence-based practice as scientific "invention" are elaborated in example 3.3, but without resurrecting discussions that we deal with elsewhere in the opening chapters and in chapter 7, evidence-based practice is often premised on a position that has been described as "an accumulative method of inductive inference that is so effective that . . . it should be given the name of 'strong inference'" (Platt 1964, 347). Platt believes that this "makes for rapid and powerful progress. For exploring the unknown, there is no faster method" (347). He does not pull his punches. Such a method provides a long-needed "absolute standard" for "maximum possible scientific effectiveness" (352).

A number of general observations may be made. First, evidence-based practice entails underlying claims regarding good practice that seem to go beyond a plea for reason-based professional judgment and practice. For example, when Sackett writes of evidence use as being "conscientious, judicious and explicit," this appears to entail three kinds of claim:

- A *moral* claim. "Conscientious" suggests done according to conscience; as a moral duty.
- A *wisdom* claim. "Judicious" suggests sound judgment and is a mark of practical wisdom and discretion.
- A claim to transparency and *openness*. "Explicit" appears to have the sense of leaving nothing merely implied.

Second, advocates of evidence-based practice often appear to distinguish hierarchies of evidence. Yeager and Roberts, for example, refer to evidence levels from meta-analysis through single randomized control trials (RCTs) to "uncontrolled trials" and finally "anecdotal designs" (2004, 6). In assessing evidence-based practice it is important to acknowledge that to almost all protagonists the hierarchies of knowledge are relative to the questions being asked and not absolute.[7] Also, other factors, such as practice wisdom and values, *are* recognized in varying ways by evidence-based practice supporters.[8] But there is still a general conviction that these should be subordinated to "rigorous consideration of current best evidence . . . of the *effects* of particular interventions" (6). Again, the more thoughtful advocates of evidence-based practice do not claim that only randomized control trials provide evidence about effects but rather that RCTs give the most secure evidence about internal validity and that that internal validity is crucial. The net effect of this is that RCTs are often believed to be the "gold standard, the Rolls Royce of evaluation approaches" in the human services (Chelimsky 1997, 101). Whether *internal* validity is quite so important is, of course, not a foregone conclusion. Cronbach's argument that " 'external validity'—validity of inferences that go beyond the data—is the crux of social action, not 'internal validity' " (Cronbach et al. 1980, 231) is explored in chapter 5.

Third, what we believe about evidence leads to corresponding assumptions about good social work practice. For example, applied to social work with offenders, the general consensus on practice implications is that practitioners should:

- Target factors that have contributed to offending
- Adopt methods that have structure and require active involvement in problem solving
- Match degree of intervention to risk of offending
- Have program integrity, that is, avoid drift, objective reversals, or noncompliance

(Davies 2000)

Fourth, there is a wider question, which I mention in passing. Commitments to certain ways of knowing typically bring with them, by their epistemological assumptions, constraints on the range of action and interventions that may follow. We develop in chapter 9 Romm's premise that "the process of attempting to 'know' about the social world already is an intervention in that world which may come to shape its constitution" (1995, 137).

Fifth, promoters of evidence-based practice within the human services are not unanimous on whether professional practice shows up positively as a result. Mullen, for example, laments that "for the most part social work practitioners are not engaged in evidence based practice" (2004, 208), and Kirk and Reid remark of social work that "within the profession, science remains on the cultural margins, struggling for a voice and a following" (2002, x). Macdonald is correct when she claims that "we frequently overestimate social work's beneficial effects and underestimate social work's capacity for adverse effects. In general, the more rigorous the research design, the less dramatic the former and the more transparent the latter" (2000b, 124).

Sixth, in the minds of some advocates of strong evidence-based practice, social workers should be bound by an ethical commitment to practice what works, professional codes of ethics should require use of empirically validated knowledge where it exists, and when they do not do so they should be open to breach of professional responsibility. Reid thought this proposal was "unlikely to see the light of day" (2001, 278), but more needs to be said. This is an individualizing position attractive to those who want to find ways to blame particular social workers for, for example, child deaths. The committed proponent of scientific practice, Geraldine Macdonald, said over twenty years ago that she "takes issue with those who assume that the responsibility for bad outcomes in social work . . . is appropriately laid at the feet of individual workers" and "argues that a morality of social work must recognize the social and organizational context in which it occurs" (1990, 525).

The question of professional accountability is reinforced by ways in which evidence-based practice is closely associated with accountability models of policy and practice. In assessing the implications of evidence-based practice, the extracts in example 6.2 illustrate how judgments about accountability are rarely straightforward. They also underscore the need for empirical studies of accountability (cf. de Kok and Widdicombe 2010).

EXAMPLE 6.2

"Just Sheer Luck" and Accountability

There remains an element of unpredictability near the heart of social work practice. For social workers in one study, whether work goes well or badly is not the product only of rational and intelligent planning. At its simplest, it is reflected in the assumption of one social worker that "there's always an element of good fortune in these things."

> Whether it was because the time was right for her and I was just lucky that I connected, I don't know. I think it was just sheer luck. It's a bit like gambling and roulette, and every now and then you score the right number and you make a good connection with somebody.
>
> I've been trying to help (the group) for three or four years. I don't know what happened. I was really dreading it, and I went in and I flew through it. I mean it was like flying, it was such a brilliant experience for me . . . It was a one-off.

The unpredictability of social work practice applies equally well to work that goes badly, although none of the practitioners used the term "bad luck" to describe it:

> You can spend a lot of time on a particular case and you feel you've done really well, only to find out that somebody rings up and tells you they're in (hospital) on the poisons unit. So, it's very hard to evaluate or to recognize whether you're doing a good job or not.

The commonly expressed belief that cases are capable of "blowing up into a total nightmare" assumes that there are aspects of work beyond the control of even the seasoned "game planner." This has implications for the *accountability* of social workers. If outcomes are no easier to predict than a number on a roulette wheel, then social workers cannot be held accountable for work going wrong or praised for work going well. Most of the social workers in this study were almost always able to identify substantial mitigating factors when work had not gone well:

> I suppose I feel there has been a failure—even if it wasn't necessarily mine.

The research elicited recurring instances of ways in which practice regarded by social workers as having gone less than well is likely to trigger deep uncertainties in the mind of the reflective practitioner. Am I responsible? Should I take the blame? Could it have been avoided? The dilemma is unforgivingly sharp. There are two realities that need to be brought into conjunction. First, the widespread sense that outcomes, both good and bad, are beyond prediction and will never cease to jump out and surprise the practitioner. Social workers—at least the ones spoken to in this study—were characterized by this uncertainty. Second, social workers display a hesitation about claiming credit for "good" work or blame for "bad" work. The two realities—predicting and accountability—are linked. In the words of one probation officer,

> I think, bearing in mind the fact that you can't predict what judges are going to do, you can't predict what magistrates are going to do, in a sense you can't take the blame for a failure. If you can't take the blame if it doesn't work, how can you take the credit if it does?

Source: Shaw and Shaw (1997).

Seventh, evidence-based practice in the United States often has had little or no sensitivity to feminist social work science. The assumption frequently made seems to be that a commitment to emancipation and a critique of patriarchy sit uncomfortably with a strong evidence-based practice position. The debate is a substantial one to which we return in some detail in chapter 8. For the moment it suffices to emphasize that the point and counterpoint are by no means settled. Take, for example, Anne Oakley (1998a, 1998b, 1999, 2014), whose books *Sex, Gender, and Society*, *Housewife*, *Subject Women*, and *The Sociology of Housework* came to be hailed as "classic" feminist texts (cf. Oakley 2014, 205). Oakley's *Experiments in Knowing* (2000) reflects her general position that "information about the effectiveness of public services is hard to come by using any other means" than rigorous evaluation methods (Oakley 1998b, 110). In one of her articles she starts with C. Wright Mills's skepticism regarding "method as a self-conscious procedure" (1999, 247) and reproduces a conversation between Mills and Paul Lazarsfeld that reportedly started with Mills saying, "Nowadays men often feel that their private lives are a series of traps," and Lazarsfeld replying, "How many men, which men, how long have they felt this way, which aspects of their private lives bother them, when do they feel free rather than trapped, what kinds of traps do they experience, etc etc?" (247–248). Oakley suggests this captures the different positions in the qualitative/quantitative argument.

Her wider commitments prompt her to comment on involving research participants and to say that randomized control trials tend not to take seriously the views of the participants. In most of the studies, "the thousands of people who took part in them appear mainly as numbers, rather than as individuals with their own stories" (1998b, 108). She enters a plea for "a more broadly based social science approach, more ethnographic and process data, the testing of simpler interventions, and more attention to the expectations of the groups studied, particularly in relation to study design." These "would all have improved the capacity of the experiments to illuminate important policy questions" (109).

The eighth and final observation regarding evidence-based practice is that it is not accidental that there is a recurring note of the elusive, "not quite yet," "but has not been completed" nature of such claims (Okpych and Yu 2014, 5). This also can be seen when Joel Fischer (1993) predicted the triumph of empirical practice by the year 2000 and occasionally in fields such as sociology. The opening words of John Madge's weighty overview of *The Origins of Scientific Sociology*, which I recall reading cover to cover as an undergraduate, were: "The theme of this book is a simple one. It is that the discipline of sociology is at last growing up and is within reach [*sic*] of attaining the status of a science" (Madge 1963, 1).[9]

Modifications, Directions, and Reservations

Responses to mainstream science-based social work practice positions have been of various forms:

- Modifications and reservations within the existing framework that "soften" the edges
- Modifications within the existing framework that reinforce the practice implications
- Responses to perceived threats to social work science from within the existing framework

Softening the Edges

The original version of evidence-based practice in the human services in the United States has gradually been re-presented in terms that suggest a need to *involve service users* and to take account of the *local contexts* in which services are delivered. General assertions that service users are—or at least ought to be—involved in decision making at the case level can be found in the recent literature. In the United States, writers who otherwise have not been strongly associated with stakeholder or participatory models of evaluation have added their voices to this claim (e.g., Corcoran and Vandiver 2004; Gibbs and Gambrill 2002; Howard, McMillen, and Pollio 2003; Mullen 2004). Similarly, the need to take account of local contexts has gradually been acknowledged. In one of the most fully developed examples of this modification, Proctor and Rosen say that "having a solid foundation of empirical support does not guarantee that a given intervention will meet the needs of a particular client" and "although sometimes given short shrift, local knowledge is an important complement to research-based knowledge" (2003, 196, 197). They warn against the risk of overcommitting to standardized interventions. This leads them to the issue of when and how an intervention should be modified to meet local contingencies. They support doing so, albeit in a careful and well-reasoned stepwise process. Practitioners should:

1. Locate evidence-based interventions relevant to the outcomes for pursuit.
2. Select the best-fitting intervention in view of the particular client problems, situation, and outcomes.
3. Supplement and modify the most appropriate and best-supported treatments, drawing on practitioner experience and knowledge.
4. Monitor and evaluate the effectiveness of the intervention.

Lying just beneath the surface is an acknowledgment that evidence-based practice has to retain a tension between routinization of practice and professional judgment. However, while this is occasionally recognized, it is not very explicit (cf. Howard, McMillen, and Pollio 2003).

These modifications are unlikely to persuade committed constructivist researchers, but they do represent a significant shift of position. There *is* a softening of strong versions of evidence-based practice on the part of some. Munson, for example, pleads for a version of evidence-based practice that is "balanced with a relationship model" (2004, 259). He argues that treatment "must have a developmental focus using a scientific perspective that relies on evidence that is grounded in a therapeutic relationship" (252). Evidence alone and relationship alone do not produce change, but both—and practice that is thus based has ethical implications. While advocates of a strong version of evidence-based practice appear to believe that the main risks of being unethical lie in being nonrational, Munson suggests that "there must be monitoring for the subtle belief that only evidence-based tasks or outcomes have value" (254). Evidence-based practice must, he believes, address power issues and ethical aspects of, for example, lack of expressive capacity in traumatized children.

Reinforcing the Implications for Practice

It would be misleading if we assumed that the only direction of evidence-based practice is toward a softening and accommodation with those who have hitherto been its critics. Considerable effort is being made to develop evidence-based practice standards (Proctor and Rosen 2003; Rosen,

Proctor, and Staudt 2003), routinized guidelines (Mullen and Bacon 2004; Okamato and LeCroy 2004; Springer, Abell, and Hudson 2002a, 2002b), and strategies for learning and teaching evidence-based practice (Howard, McMillen, and Pollio 2003). Howard and colleagues tacitly acknowledge that evidence-based practice does not provide a deliverable platform and are ready to allow that instructors "can teach interventions without compelling empirical support as long as . . . there is sufficient justification for the intervention" and the available "scientific support" for the intervention is also taught. They include "strong theoretical justification" and "practice wisdom" as grounds for teaching nonempirical practice (248).

The assumptions around the operationalizable nature of social work that lay beneath such strategies are rarely spelled out. Indeed, they often are presented as the positive and unproblematic consequences of valuing an empirically founded practice.

Responding to Perceived Threats

Rather than survey the landscape of such responses, I take a case example based on Gambrill's (2010) warnings of the dangers of propaganda in relation to evidence-based practice. I also set this out in the form of an exchange, which followed a seminar in London in 2012. I do this as a way of illustrating, albeit in a modest way, varying forms of science writing in social work. I map that question more fully in the appendix.

In example 6.3 I set out as evenhandedly as I can manage the perspectives and stance that Gambrill brings to her argument. In this and the immediately following example 6.4 I "talk" in the first and second person to the author.

EXAMPLE 6.3
Gambrill on Propaganda

First, I think I detected a recurring note of skepticism about much research "evidence" and the uses made of it by practitioners and agencies. For example, you lament "inflated claims about "what we know" about causes, about accuracy of assessment measures, about risks, and about the effectiveness of remedies" (302). You cite with approval Ioannidis as arguing that most published research findings are false and say that "conclusions often are not supported by methodologies used" and "most services are of unknown effectiveness" (303), such that there are cases of "harming in the name of helping" (304).

Second, you make a very strong plea for a shift from a medicalized frame of reference to a behavioral one, which in the course of doing so will destigmatize large areas of, for example, mental health. You decry "the relentless redefinition of problems-in-living as mental illnesses in need of help by experts" (307) and have a whole section on "disease mongering" (309–311).

Third, I detected an interesting position: that the original vision of evidence-based practice was "deeply democratic, participatory and transparent" (304) but that it has been misused to "maintain authority-based decision-making in which criteria such as consensus, anecdotal experience or tradition are used" (304). You later complain that "most clients are not involved as informed participants in making decisions" (309). I saw your stance at this point as one of seeking to return to the original vision of evidence-based practice.

Fourth, and briefly, you have what I read as a continuing faith in the power of information—"we are not free if we are not informed" (305). I read this also where you counterpoise reasoning as "true rhetoric" against "the manipulative discourse of propaganda" (304).

Fifth, I interpret your article as perhaps more about the dangers of propaganda than about evidence-based practice per se. In developing this argument you talk about a propagandist society and place evidence-based practice and social work within that wider picture. You are profoundly influenced by your reading of Ellul's work on the theme. Thus you say, "this is the technological society, dominated by the mass media" (306)—a classic Ellul position—and following Ellul you link that to the political form of society, in which "efficiency, standardization, systematization, and the elimination of variability" are dominant (307). I think you have in mind especially the United States, with the rest of the West on its coattails, as when you say Ellul "argued that the effects of propaganda are always negative (especially in a democratic society) whether intentional or not" (313).

Finally, your general response is, as I understand it, that propaganda is an inevitable part of society, so we cannot escape its influence but must aim to understand it so we can "mute its effects" (314).

Source: Gambrill (2010).

EXAMPLE 6.4

Shaw on Gambrill on Propaganda

I welcome much of what you say. For example, I endorse your efforts to suggest *standards for discourse* and your dislike for ad hominem arguments. I suspect you would like the sociologist Robert Dingwall's plea that being right is more important than being right on. You adopt Paul Grice's maxims derived from his cooperative principle for how we should talk (315), though I felt this was too undeveloped. Grice is talking about everyday situations rather than the related but particular kinds of talk that take place in academic fields.

I also like your use of Model and Rank's (1984) classification of propaganda (306) as entailing:

1. Overemphasizing the positive aspects of one's preferred model
2. Hiding and minimizing the negative aspects of one's preferred model
3. Overemphasizing the negative aspects of opposing views
4. Hiding and minimizing the positive aspects of opposing views

Your *skepticism* poses some complex issues about the relationship between power and reason. I support what I would call a Socratic skepticism against, for example, a social work culture of what we may call "false positives." But it also risks paralyzing action. This is an area where there is diversity within the broad evidence-based practice movement, as when Bill Reid asked: "Is it better to make limited but well documented progress or to work toward more important goals with less certainty of what we have attained?" (1988, 45). His choice between limited and well-documented goals or working toward more important goals with less certainty illustrates one of the central tradeoffs that appears to face practitioners, policy makers, and researchers.

Your plea for a (much) *less medicalized* approach to mental health has value and more than a grain of truth in it. It reminded me of an article about historical trends in psychiatry provision in Wales, that "it might be naïve to think we could have expected that a massive increase in the number of psychiatrists would ever lead to anything other than the treatment of more patients than ever before" (Healy et al. 2005, 40). But of course your language of "behavior" is more than a plea for demedicalizing!

However, for me, you set up an oversimplified rhetoric about the relationship between "reason" and "evidence" on the one hand (which by the way, like many in the scientific practice community, you rarely relate) and, on the other, propaganda. Despite your aside about whether propaganda is intentional or not, you do not develop any significant distinction around intentionality. You do not make connections to the extensive contemporary work on that theme, despite your plea for a sociological view. Hence you nowhere address the role of tacit knowledge. For me this all leads to an overconfidence in the powers of reason. Tied to this, I find it implausible to accept your collapsing of "consensus, anecdotal experience or tradition" into the pejorative category of "authority-based" decision making. For me skepticism is needed on the *limits* (more than limitations) of science.

Over the years there have been various criticisms of research that aspires to produce an evidence-founded, reasoned "case":

1. That in its original and best forms it is well-conceived, and our hopes for social work rest mainly therein, but that it has been badly and/or inconsistently done.
2. That the venture is ill-conceived.
3. That it is rightly conceived but that we should be much more modest in the hopes we invest in it.

You appear to take the *first* position.

Concluding Thoughts on Good Evidence

Important questions remain unresolved in discussions of evidence and social work science. What is the role of research evidence as against other sources of information, including personal experience, in policy making and professional practice? What weight should be given to evidence over against the role of judgment in decision making? What is the potential and possibility for synthesis? What different kinds of contributions can research make? How ought we to articulate commitments to evidence and to strategies of research inquiry? Hammersley sets his doubts in context of Chalmers's case for doing more harm than good:

- Can we ever prevent professionals from doing more harm than good?
- Can we determine with great certainty via research alone *whether* they are doing harm or good?

- Does evaluating policies and practices by means of research always lead to more good than harm?

(Hammersley 2005, 87)

He says that these general concerns stem from more specific ones central to which "is the sharp distinction between practitioner *opinion* and scientific research *evidence* that is built into Chalmers's argument and into other presentations of the rationale for evidence-based practice" (88).

On articulating commitments to evidence with strategies of research inquiry, traditional views of science treat "explanation" as always a question about what *causes* something. This understanding stems from the empiricist philosopher David Hume and is later mediated through Immanuel Kant. We spoke in general about the meaning and relevance of causality in chapter 1. At the heart of the question lie different root metaphors of what research is basically about. Kushner (1996) catches the different views about the core of research—for mainstream outcomes researchers and evaluators, he suggests, research is basically about *order*; for process researchers it is about *conversation*. Hence his key test of what keeps evaluators awake at night. For outcomes researchers it is not managing to distill the evaluation into a single unified story; for process researchers it is having only one story to tell.

Philip Abrams, the sociologist whose early death brought to a close a major contribution to our understanding of fields as diverse as informal care (Bulmer 1986) and historical sociology (Abrams 1968, 1982), reflected on the difficulties of experimental designs in his field.

> The resistance of informal social care to experimental evaluation has entirely to do with the problem of breaking down the intractable informality of the treatment; of reducing informal caring relationships to the sort of unit acts, factors, events, variables, [and] items needed if specifiable inputs are to be systematically related to specifiable outcomes.
>
> (Abrams 1984, 2)

"The irreducible property of informal care seems, unfortunately for social research, to be the fact that it is genuinely informal; and that it defies formalisation" (2). In consequence, research faces "the extreme difficulty of isolating inputs" and locating and isolating "treatments."

Abrams illustrates this from "Good Neighbour" schemes set up in his time to simulate the world of informal care. But "in practice, however, we are dealing with a metaphor rather than an analogy. In one all-important sense Good Neighbours are not good neighbours but something quite different" (3). The types of care offered are "fundamentally different"—indeed, Good Neighbour schemes were set up "to deliver care precisely to those people who have fallen through the net of informal care" (3). I refer to Abrams to pose a challenge regarding the nature of some forms of social life. The disjuncture is not absolute—it is possible of course to measure things such as the economic costs of care. But the problem is profound not only in intellectual terms but because it relates to the larger problems of the linkages between formal, state, or federal services and the informal sector.

Many social scientists, including some in social work, "appear to be believers in Natural Law insofar as they appear to believe that what is ethically desirable can only be desirable to the extent that it is rooted in empirical investigation; more familiarly that investigation can itself give us guidance about what we ought to do" (Nokes 1967, 79). There are Christian roots in this position (Hooykaas 1972), albeit ones that even when recognized are unlikely to be accepted. It is a shoot from a root belief that reason and ethics alike stem from the same divine order and that therefore there cannot be any inherent conflict between them. But twenty-first-century expressions of this position are potentially dangerous. The connection may appear in social work when, for instance, the assumption seems to be held that because a certain approach to intervention is a good thing, then it would be found to be effective if only it were properly tested. It is probable that the heated critical responses to 1960s intervention research, which seemed to show that social work was not demonstrably effective, were driven in part by this confusion of desirability and effectiveness. The suggestion that adherence to demonstrably effective practice should be written into sanctionable codes of ethics encounters a difficulty that to this writer appears intractable—"that only a moral argument can be advanced for some of the policies that enlightened opinion would approve" (Nokes 1967, 81). This is so because not every policy or practice aimed at the well-being, support, or change of those with or for whom practitioners work is measurably effective. The only justification left for such practices is "that they represent ways of dealing with people that we feel to be right and proper" (81). Anyone who thinks this dilemma is merely a transitional feature is living in a different universe. We have seen in chapter 4 that it is difficult to believe that technological applications of science could transform social work into a purely instrumental form of action.

There is a further problem too little appreciated. The question "Does it work?" is a skeptical question and "functions as an exclusionary gatekeeper" (Bogdan and Taylor 1994, 296). Even assuming that design problems could be solved, the question still would not be helpful to practitioners. "Conscientious practitioners do not approach their work as skeptics; they believe in what they do" (297). Bogdan and Taylor have worked to develop what they call "optimistic research." "We have evolved an approach to research that has helped us bridge the gap between the activists, on the one hand, and empirically grounded skeptical researchers, on the other" (295). The main focus of their research had been on questions of how people with severe disabilities could be integrated into the community. Rather than ask whether services are effective, they ask, "What does integration mean?" and "How can integration be accomplished?"[10]

We might concur with Nokes that to say "convictions are one thing and science quite another has really no cogency in the social sciences" (1967, 65). We have cause to turn to the relationship between moral conviction and science several times in this book, especially toward the closing chapters. Some would hold with heartfelt conviction that academic detachment is a form of dilettantism. "From this point of view, to be *engaged*, far from being a disqualification, becomes almost the primary basis of valid comment" (65). Nokes's concern is that this "is bad because it deprives whole areas of human experience, on which progressive opinion has already made up its mind, of any sort of proper appraisal" (67). He laments that this often leads to a stance where "authors rely heavily on human folly and inadequacy as an explanatory principle, the folly of society in permitting the kinds of conditions they found, and the inadequacy of people within the organization in not doing better than they did" (68).

Criminology is perhaps the social science as much as any other that is approached from a reformist position. My doctoral supervisor, Howard Jones, long ago assured us that criminology is "above all . . . a reformist study" (Jones 1962, 1). While its disciplinary entity has doubtless shifted in the intervening years, if we turn to social work the reformist position is, if anything, more marked. Edith Abbott could rarely resist poking at sociologists, so when she reported one of her major housing studies, she recorded:

> One question that immediately arose in the first house-to-house canvasses undertaken was the question whether to undertake to reform the various objectionable conditions which were found. . . . From the beginning, all our investigators were graduate students preparing for some form of social work. Unlike professional investigators or the students in a social science department in a university who are accustomed to "observation with the idle curiosity of the scientist," our investigators were accustomed to see "what could be done about it." Every effort was made to bring about an immediate improvement in any bad condition that seemed remediable.
>
> (Abbott 1936, 165)

"Experimenter's Regress"

We have seen how there is very frequently disagreement over the meaning and implications of the science of evidence in social work. There have been numerous instances in the history of social work research. One of the better documented ones was the furor that occurred following the publication of *Girls at Vocational High* (Meyer, Borgatta, and Jones 1965) and the almost immediate rejoinder from Macdonald that opened: "This book has created a storm of criticism of all social work" (1966, 175).[11] Why is social work marked by recurring controversies of this kind? In addition to the general significance of controversies, which we considered in the previous chapter, part of the grounds for such academic altercations is attributable to what Collins refers to as "experimenter's regress," which occurs in situations where the science in question is novel. "It is hard for an experiment to have an unambiguous outcome because one can never be sure whether the test has been properly conducted until one knows what the correct outcome ought to be." Hence, experimenter's regress "occurs when scientists cannot decide what the outcome of an experiment should be and therefore cannot use the outcome as a criterion of whether the experiment worked or not" (Collins and Pinch 1998, 2, 106).

One example they give is an interesting and in part moving account of efforts to assign blame for the 1986 *Challenger* explosion. Much debate centered on the O rings (in effect the seals between the sections), the existence and size of gaps between them, and whether the various tests were appropriate simulations of what happens in actual takeoff. One side argued they were too tough, the other that they were not tough enough. Here the experimenter's regress comes into play. "The 'correct' outcome can only be achieved if the experiments or tests in question have been performed competently, but a competent experiment can only be judged by its outcome" (40).

> The logic of the situation is like this. Is there a large or small gap? We need to know if there is a gap to detect. To know this we need a good test to find out. But we don't know what a good test is until we have tried it and found that it gives the correct result. But we don't know what the correct result is until we have built a good test.
>
> (40)

NASA thought their electrical measures were a good test. Thiokol, the company that supplied the O rings, thought the physical measures were a good test. NASA saw the tests as similar; Thiokol saw them as different. Given this "interpretive 'loophole' provided by the need to make similarity or difference judgments, it is not surprising that tests alone could not settle the performance of the joint" when each gave contrary results (39). Collins and Pinch give further examples, including disputes about the origin of oil and macroeconomic modeling, and Collins expounds the argument elsewhere in relation to his research on gravity waves. Collins and Pinch conclude that science and technology are far messier than they are usually presented, that conclusions and decisions about what is true are inevitably reached for reasons other than strict scientific proof, and indeed that in many cases it is difficult to reach agreement on what would constitute proof. All this implies that building knowledge is a messier business than conventional accounts imply. In seeking to resolve a dilemma of this kind we are taken back to questions of scientific argumentation (cf. Godin and Gingrass 2002).

This connects readily to the *Girls at Vocational High* exchanges, during which Macdonald responded in various ways, generally by arguing that the tests of social work effectiveness were the wrong ones. For example, "preventing school dropout, through improving grades, and forestalling illegitimate pregnancy" was to be tackled by agency workers employing individual and group counseling (Macdonald 1966, 181). She insists that the "criteria hardly seem appropriate to the program to which they were applied in the way in which they were applied" (182), whereas Meyer, Borgatta, and Jones had argued that meeting the proximate goals would lead to meeting ultimate goals. Lying just beneath this thesis and antithesis was a view about whose understanding of science and its application ought to count. As Macdonald expressed it in rather hurt terms, "for evaluative research to flourish we need a stable union between research and practice—not brief and sometimes catastrophic affairs between partners who hardly know each other" (189). Whether either "side" was substantially correct is beside the point. The "interpretive loophole" made it almost impossible for agreement to be reached.

Incidentally, Collins's argument that "experiment alone cannot settle the matter" (Collins and Pinch 1998, 87) appears as radical skepticism about the grounds of belief, though Collins never aligns himself with radical skepticism and indeed has—for example, through his ideas of expertise, through insisting that not all science works this way, and by leaving open the possibility that some forms of tacit knowledge may one day be explicable—distanced himself from the original "strong school" of the sociology of knowledge. The practical outcome of the argument about regress is that building knowledge requires (professional) judgment and social agreement about how to interpret evidence "correctly."

Positions of this kind helpfully bridge ideas of evidence and those of understanding. Schwandt, for example, intriguingly suggests that

> My presumptive knowledge counts as objective when, even though I have been as careful as I can in checking that *p* holds, I could still be surprised; I could still discover that *p* fails. . . . My belief in *p* is objective when . . . there is still something in *p* about which I might be making a mistake. My belief or knowledge is subjective when such surprises are ruled out. In our everyday life, we inhabit a world with real mistakes and surprises built in, and, hence, objective knowledge built in.
>
> (Schwandt 1999, 459)

We now turn to these issues, along with questions of how expertise and tacit knowledge play important roles in social work science, in chapter 7. They will lead to more fundamental questions regarding evidence and science. Foucault argued, speaking of the origins of clinical medicine in France, that it was not new and more sophisticated insight but a change in how "illness" and the "doctor" were defined—"the new doctor started looking in a different way at a differently constructed object of scientific knowledge, namely illness" (O'Farrell 2005, 37). His focus was not only on how an object (for example, madness) is constructed but how knowers (in this case, doctors) are constructed. Or, as we might say, echoing part of the field covered in chapter 3, it is not only how "evidence" is constructed but also how "evidencers" (social workers) are constructed.

Taking It Further

Reading

Kirk, S., and W. J. Reid. 2002. *Science and Social Work*. New York: Columbia University Press.
Graybeal, C. T. 2007. "Evidence for the Art of Social Work." *Families in Society* 88(4): 513–523.

For a thoughtful if basically accepting account of forms of evidence-informed social work Kirk and Reid (chaps. 4, 5, and 7) is a good introduction. But the tendency to take the United States as a template for the world, and the general confidence in science, mean that it needs supplementing with other reading. Graybeal would be one such supplement.

Hammersley, M. 2005. "Is the Evidence-Based Practice Movement Doing More Harm Than Good? Reflections on Iain Chalmers's Case for Research-Based Policy Making and Practice." *Evidence and Policy* 1(1): 85–100.
Hammersley, M., and R. Gomm. 1997. "Bias in Social Research." *Sociological Research Online* 2(1). http://www.socresonline.org.uk/socresonline/2/1/2.html/

The two Hammersley pieces should be read together and not as alternatives. Hammersley helpfully has got under the intellectual skin of authors as different as Iain Chalmers and Norman Denzin. If these are found rewarding, then readers will find his sally into social work worth checking out:

Hammersley, M. 2003. "Social Research Today: Some Dilemmas and Distinctions." *Qualitative Social Work* 2(1): 25–44.

Task

Members of the social work community tend to hold definite views regarding the merits or otherwise of evidence-based practice.

- Take a research article that you believe shows the plausibility of the general position you hold in this regard.
- Review the themes and issues raised in this chapter and set out reasons why your general position may be less strong than you presently believe.

[7]

Social Work Science and Understanding

In this chapter I continue the exploration of three central concerns for social work science—evidence, justice, and, here, understanding. The challenges and questions are, if anything, still more extensive and less tractable than concerns about evidence in social work science. In the first part of the chapter I consider what is intended when we speak of understanding and, in certain senses, knowing something. I then ask:

1. In what senses do social workers and social work scientists possess expertise?
2. In what ways does it make sense to speak of social work science as shaped by commonsense knowledge?
3. What do we know explicitly or tacitly?
4. What is involved when critical reflection, judgment, and discretion are exercised in social work science?
5. In what ways can we say that choices regarding method in social work inquiry are formed by science as understanding?

Understanding

By way of anticipation, this first part of the chapter considers what is intended when we speak of understanding and, in certain senses, knowing something. This will pick up leitmotifs already touched on in the opening chapters and create an agenda for later parts of this chapter. To understand and to know (in the particular senses of having personal experience of or familiarity with something) generally carry positive associations. In a more specialist sense, "understanding" is something to which social work often seems to make special claims. "The assumption that understanding people is difficult forms one of the assumptions that distinguish modern social work from earlier benevolence" (Timms and Timms 1977, 135). In a now superseded set of requirements for social work training in England the following requirements were made:

> In your specific area of practice, you must understand, critically analyse, evaluate and apply the following knowledge:
>
> 1. The legal, social, economic and ecological context of social work practice.
> 2. The context of social work practice for this area.
> 3. Values and ethics.
> 4. Social work theories, models and methods for working with individuals, families, carers, groups and communities.[1]

The relevant point for our purposes is that there are different kinds of knowledge and understanding in this list. To understand social work theories or the legal context of work in a given area of practice is not the same as to understand and apply values or methods of working. To express this differently, knowing that something is the case is not the same as understanding *how* to do something.

Noel Timms foregrounded understanding in ways rather distinct from standard social work treatments when he confessed his personal "delineation of social work as concerned with understanding rather than information, and understanding not necessarily with a 'practical' or predetermined end in view. . . . We should consider social work as primarily neither an applied science nor simple good works but a kind of practical philosophising" (1972, 2–3). Understanding has, however, an ambiguous and often uncomfortable relation with social work science. This is so for at least two general reasons. First, to understand is associated with inner kinds of knowledge and with experience, with their immediate apparent tension with scientific knowledge. Second, understanding and experience are not thought of as the special privilege or domain of science, and hence the question arises of the relationship between scientific knowledge and "lay" knowledge, expertise, or common sense. We will take a preliminary look at these two questions and return further to questions of expertise and common sense later in the chapter.

Understanding and Experience

A distinction often drawn is that between what we derive from our senses as distinct from knowledge by description. A couple of examples may serve to show how this way of considering the world—between "internalist" and "externalist" ways of justifying what we believe—is deeply rooted in social work and cognate fields, usually with the apparent belief that the one is in tension with the other. The evaluation writers Stake and Trumbull may seem to suggest as much when they argue that "for practitioners . . . formal knowledge is not necessarily a stepping stone to improved practice. . . . We maintain that practice is guided far more by personal knowings" (1982, 5). The phrase "experts by experience" has sometimes been used by service users. For example, Beresford says, in terms that may overclaim, that "what distinguishes service user knowledge (or knowledges) and what is unique about it, is that it is based on direct experience" (2010, 12). Carl Rogers's classic book on *Client-Centered Therapy* spoke in a similar way when he wrote that "diagnosis is a process which goes on in the experience of the client, rather than in the intellect of the clinician." He drew the inference that "in order for behaviour to change, a change in perception must be *experienced*. Intellectual knowledge cannot substitute for this" (1965, 222). He goes further, cautioning that diagnosis engenders the feeling that self-knowledge depends on an outside expert, which in turn fosters a situation where the many are controlled by the few.

To take a position in relation to such debates the terms need to be understood. *Experience* comes from an earlier word, *empirical*. "Empirical and the related empiricism are now in some contexts among the most difficult words in the English language" (Williams 1983, 115). Empirical came into English from roots in Greek, with senses of "experience," "skill," and trial or experiment, thus showing its complex semantic potential. Its modern use is linked to "exceptionally complex philosophical and scientific movements" (115), though its simplest usage today is to indicate a reliance on observed experience—"but almost everything depends on how experience is understood" (116). It retains the tension between knowledge from observation (experience or experiment) and knowledge from directing principles or ideas. This is, of course, closely related to the more common distinction between practical and theoretical.

These issues are further complicated by a philosophical argument about the relative contributions of reason and experience to how ideas are formed. Do good ideas in social work science come from within science or from outside? My general view is the latter. To draw again on the experience of leading people in the mainstream sciences, Murray Gell-Mann, a Nobel-winning theoretical physicist, recalled a seminar more than twenty years previously.

> There were two painters, a poet, a biologist and one theoretical physicist—me. We were all talking about occasions on which we had gotten creative, useful ideas in our different fields, and it was remarkable that our descriptions were all very nearly isomorphic. In each case there was a difficulty in the way, and in each case the person thought very hard about how to overcome that difficulty. . . . They each initially had been unable to do so . . . and yet outside of conscious awareness, the mind seemed somehow to go on working on the problem, because at some point later on—say while running or cycling or shaving or cooking—the person would suddenly get a useful idea about how to overcome the difficulty.
>
> (Wolpert and Richards 1997, 162–163)

This is reminiscent of Max Weber's 1919 remark that "ideas occur to us when they please, not when it pleases us." For example, "as Helmholtz states of himself with scientific exactitude: when taking a walk on a slowly ascending street. . . . Ideas come when we do not expect them, and not when we are brooding and searching at our desks" (1948, 136).

We might hear someone say a research report is "crudely empirical," meaning it lacks any controlling principles or ideas, or that a research report is "empirically convincing," meaning that the knowledge is reliable or that an argument has been convincingly presented from the data. But to use the terms—as we do so often—as simply countering praise or blame words just confuses the issues. When we hear of, for example, "Anglo-Saxon empiricism," to distinguish European and North American general positions, "the argument usually goes beyond serious reach" (Williams 1983, 117).

Experience has two main senses:

1. "Knowledge gathered from past events, whether by conscious observation or by consideration and reflection" (126), as when we talk, for example, of the lessons gained from experience.
2. "A particular kind of consciousness, which can in some contexts be distinguished from 'reason' or 'knowledge'" (126). This speaks of experience of the *present*, involving a full, active kind of consciousness or awareness.

Experience in the second sense seems to call for an appeal to the whole consciousness as against relying on restricted specialized faculties. This appeal to "wholeness" is common in social work and obviously stands for an appeal to "inner" knowledge and against approaches that are seen as excluding certain kinds of consciousness as "subjective," "emotional," or "personal." Arguments for reflective practice in social work seem to have elements of both past and present in different contexts, though both refer to a contrast with the kind of consciousness involved in reasoning and conscious experiments. Each sense is very different, but they share an opposition to starting from presuppositions. The two senses are not wholly apart—"in the deepest sense of experience all kinds of evidence and its consideration should be tried" (129).

This area will remain a fundamental controversy for social work, but it is often confused by failing to engage with the key distinctions. Once we express knowledge by acquaintance in language, it becomes knowledge by description. The face-to-face interview between client and social worker entails knowledge by acquaintance on both sides, but, as soon as the social worker refers to that interview verbally or in writing (in a record, e-mail, or report), it becomes knowledge by description. Likewise with a research article or report: "It is a deep question how we learn to name objects of common experience . . . from our private knowledge by acquaintance" (Gregory 1987, 412). Hence the acquaintance/description distinction has limited purchase in practical terms, in that it is now widely accepted that there is description within perception and experience.

Stake and Schwandt (2006) have explored a different, though related, distinction—that between experience-far and experience-near knowledge. In their discussion of discerning quality in evaluation, they distinguish between "quality-as-measured" and "quality-as-experienced" (we might say "knowledge" rather than "quality"). In the first case, science takes on the characteristic of "thinking criterially" through explicit comparison to a set of standards—is it reliable, valid, free from bias, and so on. Judging criterially is more or less an experience-distant undertaking. "Knowledge-as-experienced" starts from the view that scientific knowledge is a phenomenon we personally experience and only later make technical, if need be: "This view emphasizes grasping quality in experience-near understandings, that is, in the language and embodied action of those who actually are undergoing the experience of a program or policy. Criterial thinking is important but it is rooted in interpretation of personal experience" (Stake and Schwandt 2006, 408). The idea of "practice-near research" (e.g., Froggatt and Briggs 2012, Hingley-Jones 2009, White et al. 2009) explicitly draws on this way of distinguishing.

Understanding and Common Sense

Writing in 1974, when "many of our ablest and most dedicated graduate students are increasingly opting for the qualitative, humanistic mode," Campbell expounded his insistence that qualitative, commonsense knowing is the building block and test of quantitative knowing (1979, 49), seeming to suggest an articulated role for commonsense knowing at early and late stages of scientific work. More generally, Schutz took the position that "the thought objects constructed by the social scientists refer to and are founded upon the thought objects constructed by the common-sense thought of man living his everyday life among his fellow-men" (1967a, 6). This is what Schutz means when he refers to "second-degree" (sometimes second-order) constructs, or constructs of the constructs.

Some anthropologists, such as Clifford Geertz, have suggested several reasons why we should treat common sense as a cultural system—"a relatively organized body of thought, rather than just what anyone clothed and in his right mind knows" (1983, 75). If it *is* a cultural system and not mere matter-of-fact apprehension of reality, then "there is an ingenerate order to it, capable of being empirically uncovered and conceptually formulated" (92). It is here, of course, that the argument "bites" on social work science.

Pleas to treat common sense seriously must in part include pleas to treat it empirically. Commonsense pictures of the odd, the deviant, the tediously mundane, and the difficult provide us with out-of-way cases, and are hence of use to us, "by setting nearby cases in an altered context" (92).

Geertz undertakes this "disaggregation of a half examined concept" (93). He contends that the uses we gain from common sense are by understanding its "stylistic features, the marks of attitude that give it its peculiar stamp" (85), rather than its varied content. He identifies properties of "naturalness," "practicalness," "thinness," "immethodicalness," and "accessibleness" as those general attributes of common sense found everywhere as a cultural form.

Common sense is, however, not to be seen as a way of sense making characteristic only of untutored citizens, which then are explained by the professional or academic expert. The assumption that researchers possess expert knowledge different from and inherently superior to "citizen science" has been widely questioned. The conclusion has been widely reached that "the products of systematic inquiry will not necessarily be better than the presuppositions built into traditional ways of doing things. It is a modernist fallacy to assume otherwise" (Hammersley 1993, 438). Recent reflections on the place of professional practice wisdom owe an indirect debt to this realization. If social work research is to be relevant, it cannot ignore the commonsense ways in which practitioners endeavor to make evaluative sense of their practical activities. For example, Kirk and Reid (2002) argued that we know little of the everyday epistemologies of practitioners.

In this connection, with Alison Heawood-Shaw I have suggested that social workers have two models of evaluation in their heads: "evaluation proper" ("evaluation with a capital *P*") and informal, often tacit, self- and peer-oriented evaluation. It is the latter that is embedded in their practice. For instance, evaluation strategies were described as rather like having a "game plan," within which success is untoward and contingent in part on "sheer luck." Social workers judged their practice according to whether:

- Their work produced emotional rewards
- The case was "moving"
- Intervention won steady, incremental change
- Practice was accomplished without inflicting harm through the operations of the welfare system
- Confirming evidence was available from their fellows

"They were preoccupied with causes and reasons, held strongly worked views about the complexity and ambiguity of social work practice, and were sensitive to the constant interplay of knowing and feeling in their practice" (Shaw and Shaw 1997a, 1997b).

The renewed influence of Aristotelian views of theory and practice has had two main effects. First, it has led to a welcome reinstatement of the ethical dimension of reasoning and practice. Second, it has rescued notions of the practical from its status as second tier, derivative, and derived prescriptively from formal theory. For example, Schwab's influential paper on understanding curriculum argued for a language of the practical rather than the theoretical. By practical "I do *not* mean . . . the easily achieved, familiar goals which can be reached by familiar means" (1969, 1) but rather "a complex discipline . . . concerned with choice and action" (1–2).

We opened this section by linking understanding and certain meanings of knowing. But in more usual usages understanding and knowing are related at a greater distance. "To *know* is to signal that we have engaged in conscious deliberation and can demonstrate, show or clearly prove or support a claim" (Schwandt 1999, 452), whereas to understand entails what we make of something. We "are always engaged in trying to 'make something of that' " (453), as Geertz observed regarding the Moroccan sheep-stealing story that illuminates his often cited but perhaps less often read essay on thick description. The task at hand is "figuring out what all that rigmarole with the sheep is about" (Geertz 1973, 18). We come to understand as part of practical and social life, rather than as a private, interior undertaking, such that understanding and interpretation are practical-moral activities and have to do with entering a dialogue. As Schwandt remarks, "the act of understanding is existential, not merely, or even, exegetical" (1999, 457). Unlike the interpretive relativism of some commentators, the very effort to make sense of a matter presupposes that we can be wrong. In that respect the position taken here and through this book may be set over against some postmodern movements. To quote Geertz in remarks made near the end of his career: "People can be understood. . . . It is difficult, it is hard, and the differences are real: they are not to be papered over by some sort of hypergeneralized description of things that buries all the differences. But it is possible to understand people across very large gulfs of difference."[2]

Common Sense, Expertise, Tacit Knowledge, and Science

Common Sense and Science

Our knowledge of the world "is not my private affair, but from the outset intersubjective or socialized," remarks Schutz in his essay on common sense and scientific interpretation of human action (1967a, 11). In my everyday world "I take it for granted that intelligent fellow-men[3] exist." But I take it for granted also that the same incidents and objects in my everyday world will mean something different to me and to those around me. We have different purposes and respective "systems of relevances" (11), but we overcome such differences by assuming that they are exchangeable or congruent. Thus we assume a "common sense" and reciprocal perspectives—what Weber called "of-course statements." However, "the knowledge of the man who acts and thinks within the world of his daily life is not homogenous; it is (1) incoherent, (2) only partially clear, and (3) not at all free from contradictions" (Schutz 1967c, 93). For example, it is incoherent because we select what is relevant to know more about on basis of interests that are not themselves coherent—our "interests are shifted continually" (94).

In a practice that may eventually become archaic, putting a letter in the mailbox, "I expect that unknown people called postmen, will act in a typical way, not quite intelligible to me, with the result that my letter will reach the addressee within typically reasonable time" (17). Expressed more generally, we are working with what Schutz calls *course of action types* and *personal types*. But in doing so "we can never grasp the individual uniqueness of our fellow-man in his unique biographical situation." Recalling his distinction between the "in-order-to-be" motive and "(genuine) because-motives" (see chapter 2), how then might we speak meaningfully of rational action within commonsense experience?

> We come, therefore, to the conclusion that "rational action" on the commonsense level is always action within an unquestioned and undetermined frame of constructs of typicalities of the setting, the motives, the means and ends, the courses of action and personalities involved and taken for granted. . . . Thus we may say that on this level actions are at best partly rational and that rationality has many degrees.
>
> (33)

Assumptions of shared rationality "will never reach 'empirical certainty' . . . but will always bear the character of plausibility, that is, of subjective likelihood (in contradistinction to mathematical probability). We always have to 'take chances' and 'run risks'" (33).

In his essay on "the well-informed citizen" Schutz suggests we can distinguish among "the expert, the man on the street, and the well-informed citizen" (1967b, 122). As we move between different provinces of knowledge at different times we are one or all of these, according to the interests and knowledge we bring. In this the relationship between interest and social distribution of knowledge is central. Our interest breaks down things into relevances. Also in our different (but interpenetrating) roles (for example, as researcher, program volunteer, member of a faith community, or neighbor) our interests shift and even may be incompatible. A social work application of this argument about shifting relevances and expertise can be seen in example 7.1, by considering what takes place when someone moves, for example, from being a social work practitioner to holding a faculty position on a social work program and embarks on a process of occupational socialization in the academy.[4]

EXAMPLE 7.1
Occupational Expertise

Becker and Carper developed the idea of "commitment" some years ago as a way of understanding occupational identity and career. It seems a plausible assumption that occupational identity for social work faculty and researchers will not be the same as that for practitioners. This simple notion points to an understanding of one reason why research/practice relations in social work are difficult—that is, that separate occupational cultures exist and indeed become established as entailing a set of contrasting commitments. Occupational titles, ideologies, the nature of tasks, and the kinds of organizations differ greatly for social work practitioners and academic faculty.

Moreover, how practice and research are seen in social work should not be seen in a static way. Changes in identity occur via various mechanisms. These include "the development of problem interest and pride in new skills, the acquisition of professional ideology, investment, the internalization of motives, and sponsorship" (Becker and Carper 1956/1970, 198). We also need to take into account that such change may produce *conflict* where an individual ends up with either one element of occupational identity incongruent with others or indeed the occupational identity in conflict with, for example, family and wider social identity (Becker and Carper 1957/1970).

We can reasonably hypothesize that making a commitment to engaging with research has costs for a social worker in the academy—it may violate expectations of occupational community and lead to occupation change, with risks of abandoning earlier career "investments." It may entail costs such as loss of status or social network connections. The adjustment in career interests also may entail loss through unfitting oneself for other future positions. Becker takes from Goffman's analysis of face-to-face interaction the idea that a front—a "safe face" developed to support one identity—may be lost in the shift to a new identity. In social work, for example, a commitment to advocacy on behalf of service users may be lost in research situations where stances of impartiality may be valued and expected. Finally, in social work the move to a university typically occurs after a lengthy period in practice, where such commitments have become established, and hence pressure to continue to act consistently with those will be considerable (cf. Becker 1960/1970, 285).

Why are shifts of the kind described in example 7.1 often experienced as difficult? Understanding this experience can be understood partly in terms of ways in which one may be an insider or an outsider and similarly in the light of what it means socially to be a stranger. We develop those arguments in the following chapter.

Where does this leave us regarding the relationship between common sense and science? As a way into these questions, we may start from the discussion earlier in the book about the relationship between the natural and social sciences. In Giddens's exposition of the idea of the double hermeneutic he takes the position that the natural sciences share with the social sciences a concern with hermeneutics and that "the old differentiation between *Verstehen* and *Erklären* has become problematic" (1993, 9). But this does not dissolve the difference between the social and natural sciences because when social scientists, including social work scientists, seek to describe "what someone is doing" in any given context, this means knowing what the agents themselves know. It depends on what Giddens calls "mutual knowledge" shared by social scientists and the participants.

There is no parallel in the natural sciences, "which [do] not deal with knowledgeable agents in such a way. . . . Lay actors are concept-bearing beings, whose concepts enter constitutively into what they do; the concepts of social science cannot be kept insulated from their potential appropriation and incorporation within everyday action" (9).[5] By contrast, the "relation between the natural scientist and his or her field of investigation . . . is neither constituted nor mediated by mutual knowledge" (14). As he succinctly expresses it, in words where social work science may be substituted for sociology,

> Sociology, unlike natural science, stands in a subject-subject relation to its "field of study," not a subject-object relation; it deals with a pre-interpreted world, in which the meanings developed by active subjects actually enter into the constitution of production of that world; the construction of social theory thus involves a double hermeneutic that has no parallel elsewhere.
>
> (154)

How does this argument contribute to thinking about the possible relations between science, common sense, and expertise? Giddens's extended line of reasoning is one that social work scientists may—I believe should—find congenial. By mutual knowledge he means "the interpretive schemes whereby actors constitute and understand social life as meaningful," and by common sense a "more or less articulated body of theoretical knowledge, drawn upon to explain why things are as they are, or happen as they do." Such beliefs typically underpin mutual knowledge and give mutual knowledge "ontological security" (121). Common sense is by no means solely practical in character but also includes understanding of expert knowledge.

An example of how lay and expert knowledge interact is given by Williams and Popay's account of how the intervention of lay knowledge poses a challenge to the dominance of the medical profession, though "the relationship of this challenge to the world of expertise is an ambiguous one" (1994, 119). Lay knowledge presents two challenges. First, it represents a challenge to the "objectivity" of expert knowledge. "In this sense it provides an *epistemological* challenge to expert knowledge" (120). Second, "lay knowledge represents a challenge to the authority of professionals to determine the way in which problems are defined in the policy arena. In this sense it is a *political* challenge to the institutional power of expert knowledge in general, and medical knowledge in particular" (120).

We take up the second challenge in chapter 8, but with regard to the epistemological challenge to expert knowledge, "medical knowledge," for example, "rests upon the concept of disease; lay knowledge is rooted in the experience of illness" (120). By analogy we might say that social work knowledge rests upon concepts of intervention outcomes and various notions, however they are named, of persons in an environment, whereas service user knowledge is rooted in the experience of disadvantage. Though the patient's view was a focus of interest in the early twentieth century, it was "never seen as a form of knowledge in itself, but only as evidence of, or a window upon, pathology" (121). The parallel with social work needs no stating. Yet studies of lay knowledge in medicine suggests that it is:

1. Many and varied
2. Logically consistent and coherent, even when based on strong emotions
3. Biographical, consisting of "narrative reconstructions of the relationship between the illness and the person's life conceived as a whole" (122)
4. Culturally framed within systems of belief and action so in that sense "representative"

This does not justify romanticizing lay knowledge. It is worth mentioning, if only as a cautionary tale, Lindsay Prior's (2003) work on the flaws in lay understandings of illness, particularly their causes, as opposed to the experience of being ill.

EXAMPLE 7.2
On the "Causes" of Arthritis

The medical sociologist Gareth Williams, in his classic study of the genesis of chronic illness, asked a deceptively simple question: "Why do you think you got arthritis?" This led to narratives that addressed the genesis—we may say "causes"—of their arthritis, in ways that were very different from the medical mainline account of hereditary predispositions. He might have asked "how did you get arthritis?" with perhaps different sorts of answers. The question, as asked, prompted narrative reconstructions of their biographies.

Bill—*narrative reconstruction as political criticism*: "I didn't associate it with anything to do with the works at the time, but I think it was chemically induced. I worked with a lot of chemicals, acetone and what have you. We washed our hands in it, we had cuts, and we absorbed it. Now, I tell you this because it seems to be related. The men that I worked with who are all much older than me—there was a crew of sixteen and two survived, myself and the gaffer that was then—and they all complained of the same thing, you know, their hands started to puff up. It seems very odd."

Gill—*narrative reconstruction as social psychological*: "Well, if you live in your own body for a long time, you're a fool if you don't take note of what is happening to it. I think that you can make naïve diagnoses which are quite wrong. But I think that at the back of your head, certainly at the back of my head, I have feelings that this is so and that is so, and I'm quite certain that it was stress that precipitated this. Not simply the stress of events that happened but the stress perhaps of suppressing myself while I was a mother and wife; not 'women's libby' but there comes a time in your life when you think, you know, 'where have I got to? There's nothing left of me.' "

Betty—*narrative construction as theodicy*: "I've got the wonderful thing of having the Lord in my life. I've got such richness, shall I say, such meaning. I've found the meaning of life, that's the way I look at it. My meaning is that I've found the joy in this life, and therefore for me to go through anything, it doesn't matter really, in one way, because I reckon that they are testing times. . . . You see. He never says that you won't have these things. He doesn't promise us that we won't have them. He doesn't say that. But He comes with us through these things and helps us to bear them and that's the most marvelous thing of all."

Source: Williams (1984).

It appears from these accounts in example 7.2 that lay perspectives incorporate different notions of cause. Rather than cause of the disease in a general, uniform sense, lay persons are preoccupied with causes of disease in relation to the person's experience of its impact. For example: "Why has this happened to me? Why has it happened now?" (Williams and Popay 1994, 123). This illustrates a point Witkin makes: judgments about cause are influenced by, for example, whether we think that someone could or ought to have acted differently. Thus "causes are not so much discovered as negotiated" (2001, 199).

Giddens reflects on the relationship between hermeneutics and rationality. Take the belief held by alien cultures that a human being may at the same time be a raven. Is this an example of "prelogical" primitive thought because it does not recognize the principle of contradiction? He says such a belief is not obviously different from beliefs that, for example, increasing velocity lengthens the passing of time or from some versions of the Roman Catholic doctrine of transubstantiation that bread is at same time the body of Christ. His response is that mediation between different frames of meaning cannot be treated as if formal logic applies. "Formal logic does not deal in metaphor, irony, sarcasm, deliberate contradiction and other subtleties of language as practical activity" (154). For example, we might say, "It's raining. I don't believe it," without it being a contradiction. Schutz remarks to similar effect in his essay on "The Stranger." In daily life we are only partially interested in the clarity of what we know about something. And we may hold statements that are inconsistent with one another without necessarily being fallacious. Schutz suggests that to understand this we would need a "logic of everyday thinking," but "up till now the science of logic has primarily dealt with the logic of science" (1967c, 95).

Giddens extends this to scientific language. He is not saying that this dispenses with notions of contradiction but that how this is expressed always has to be grasped contextually. Consider how a person with schizophrenia talks. From a behavioral position such talk may be seen as not "authentic," but if, as might be claimed, schizophrenic talk is a transposed form of ordinary speech, it can be understood as authentic, "thus establishing the possibility of dialogue between schizophrenic and therapist" (155).

However, we should not regard the relation between technical vocabularies of social science and lay concepts as fixed. Just as social scientists adopt everyday concepts and use them in specialized ways, so also lay actors "tend to take over the concepts and theories of the social sciences and embody them as constitutive elements in the rationalization of their own conduct" (167). Because of this "the logical status of generalizations is in a very significant way distinct from that of natural scientific laws" (154). Causal generalizations, although they may in some ways resemble natural scientific laws, have an essential difference—once announced they "are picked up as such by those to whose conduct they apply, and their form is altered" (167). This has a consequence for how we ought to think about the uses of science, in that original ideas and findings in the social sciences tend to 'disappear' to the degree to which they are incorporated within the familiar components of practical activities. This is one of the main reasons why social science does not have parallel 'technological' applications to natural science" (15).

Expertise

The porous nature of the relationship between science and forms of lay knowledge obviously makes any hermetically sealed distinction between common sense and expertise untenable. Nowotny is not the only one to ponder the implications of a sentence beginning, "If we are all experts now" (2003, 156). Schutz also speaks in connection with commonsense thinking of a state of "wide-awakeness" as "a plane of consciousness of highest tension originating in an attitude of full attention to life and its requirements" (1967d, 213). Much social work science has offered superficial and sometimes misleading understandings of expertise.

Social workers often face the dominance of social science and research "experts" over practice "beneficiaries." As Schön has expressed it: "Research and practice are presumed to be linked by an exchange in which researchers offer theories and techniques applicable to practice problems, and practitioners, in return, give researchers new problems to work on and practical tests of the utility of research results" (1992, 53). Contrary to this, research and practice need linking in ways that release the potential for practice to challenge social work science and in so doing contest conventional hierarchical ways of seeing expert/beneficiary relationships.

If it is no longer clear that scientists and technologists have incorrigible access to the truth, why should their advice be specially valued? There has certainly been some rowing back from a strong position on the relativity of scientific expertise (cf. McClean and Shaw 2005, Schmidt 1993), yet it continues to be much more difficult to separate the credentialized social work researcher/scholar from the experienced practitioner or user than was once thought. The sociologists of knowledge Collins and Evans have made perhaps the most influential contribution to this general field, albeit one almost completely bypassed in social work (e.g., Collins 2010, 2014;

Collins and Evans 2002, 2007). Collins has been working on three fields: expertise, tacit knowledge, and actions. In an influential discussion paper on studies of expertise and experience, Collins and Evans (2002) appear in some ways to reinstate the distinction between the scientific community and the "laity," but without making that boundary coterminous with the boundary marking the possession or absence of expertise. For them this is not the interesting distinction. Rather the interesting distinction is between those with relevant expertise and those without, and, crucially, it is this boundary that does not map onto the distinction between science and laity. They offered a provisional classification of expertise that has value for social work and that they claim helps understand "the *pockets of expertise* among the citizenry" and so "help put citizens' expertise in proper perspective alongside scientists' expertise" (250, 251).

They adopt the term "experience-based experts," which almost exactly mirrors the term "experts-by-experience" that is often adopted by service user researchers. They "abandon the oxymoron 'lay expertise.' . . . Those referred to by some . . . as 'lay experts' are just plain 'experts'—albeit their expertise has not been recognized by certification; crucially they are not spread throughout the population, but are found in small specialist groups" (238). Collins and Evans caution that though the phrase "experience-based experts" shows the importance of experience, "experience, however, cannot be the defining criterion of expertise. It may be necessary to have experience in order to have experience-based expertise, but it is not sufficient" (252). While this leaves open the question of what kinds of experience are relevant, they offer a distinction between three levels of expertise (255):

- No expertise
- Interactional expertise: fluency in the language of a domain that need not be accompanied by any practical expertise—this requires skills related to linguistic interaction
- Contributory expertise: "being able to contribute fully to the life of the relevant community" (Evans and Plows 2007, 830)

There are obvious boundary and definitional questions with this classification, though in Collins and Evans's view they do not have to be fatal. There is a partial parallel with Schutz's distinction between the expert, the man on the street, and the well-informed citizen. The thrust of their argument is that there are groups of different kinds of experts in both the scientific and the "lay" communities. In this they echo the view of most mainstream researchers and practitioners that some sort of balance of expertise is required. They believe that one can also talk of "referred expertise" when "having expertise in one domain can assist in making judgements about a cognate or related domain" (Evans and Plows 2007, 832). There is also a kind of expertise that consists in being able to see that someone has made a mistake. These distinctions open out how we may understand the idea of "experts by experience" such that we ask questions such as: What kind of expertise is that? How is it achieved? How is it used without mistakes? Does it bring any referred expertise?

While we may think that it is contributory expertise, claims are sometimes wider and include some kind of claim of referred expertise (for example, that service users in the mental health field may have expertise on the experience of service users who are young carers), though whether this is contributory or interactional expertise is less clear. By elaborating in these kinds of ways it enables us "to ask more precise questions about the kinds of knowledge that experts and non-experts possess" (832).

The idea of interactional expertise connects with Schutz's elaboration of contextual relevances. Expertise regarding social work science should not be treated as synonymous with scientific knowledge. The Swiss writer on the relationship between science and society Helga Nowotny describes expertise as "transgressive" (2003, 152). She has two considerations in mind. "First, it must address issues that can never be reduced to the purely scientific and purely technical. The issues expertise confronts, the practices that are to be analysed and assessed as to their consequences, are characterised by overlaps and inter-linkages that bind the specialised scientific knowledge to its local and societal context" (152). Witkin says something similar for social work. "Noticing involves extracting something from its context." It entails "an intertwined, interdependent relationship between what is noticed and various contexts. Intelligibility depends on the contexts assumed, their compatibility with culturally recognized explanations, and understandings of what is possible within these contexts" (2000, 102, 103).

The second sense in which Nowotny regards expertise as transgressive "is that it addresses audiences that are never solely composed of fellow-experts" (152). In both these respects expertise connects with questions of judgment in science, questions to which we will return subsequently.

Tacit Knowledge and Expertise

Social work attention to tacit knowledge has been intermittent and rather slight in substance. It has not gotten to grips with the literature on this field and lacks empirical work, relying more on trying to work out arguments from basic standpoints. We touched on personal and tacit knowledge in chapter 2, in particular in relation to Michael Polanyi's (1966) classic text, where he sets out his understanding of the relationship between personal and scientific knowledge, concluding that if science seeks to eliminate all personal elements of knowledge "the ideal of exact science would turn out to be fundamentally misleading and possibly a source of devastating fallacies" and that "the process of formalizing all knowledge to the exclusion of any tacit knowledge is self-defeating" (20).

Polanyi has been little noticed in social work. A rare exception is Ruth Dean's (1989) essay "Ways of Knowing in Clinical Practice." However, she is taking a position quite different from that held by Polanyi, in that she extracts the idea of personal knowledge and uses it as the basis for a relativistic view of knowledge in clinical practice. Though Polanyi rejected logical positivism, he did not align himself with "ordinary language" philosophy either. Amartya Sen believes the main reason is that Polanyi *chose* to remain outside mainstream philosophy. He had "little patience for finicky discussions in professional philosophy" (foreword to 2009 reprint of Polanyi, xvi). There is also an occasional tendency to capitalize on discussions of tacit knowledge to bemoan the loss of creativity through homogenization and standardization. Martinez-Brawley and Zorita, for example, lament that social work education and practice have moved from "preparation for a cause or an artistic undertaking" based primarily on tacit knowledge to "preparation for a technological undertaking or market endeavour which requires codified knowledge" (2007, 534). This echoes some criticisms of the information society that we observed in chapter 4. One problem with each of these social work accounts is that they tend to take the meaning of "tacit" as obvious and without need of explication.

Collins (2010) has developed an extended argument for understanding different forms of tacit knowledge and for seeing them in relation to explicit knowledge. He is really dealing with a different set of issues than Polanyi, for whom it is central to how he sees wider moral and political movements in society, and the latter has a historical dimension absent from Collins. There is, Collins suggests, a problem with the term "tacit," arising from a tension between tacit as meaning "is not explicit" and tacit as meaning "cannot be made explicit." Indeed, social work accounts generally slide over this distinction, though the underlying assumption of most social work discussions seems to be that "in principle, with enough effort, any piece of it could be rendered explicit" (Collins 2010, 11) and "unravelled" (Imre 1985, Martinez-Brawley and Zorita 2007, Pawson et al. 2003, Zeira and Rosen 2000).

In the first case, the antonym of "tacit" is "explicit"; in the second case, the opposite is "explicable." Collins takes issue with Polanyi saying tacit knowledge is to "know more than we can tell" (Polanyi 1966, 4). The key words "know," "tell," and "can" ("cannot") all are in need of explication.[6] When Polanyi says that "all knowledge is either tacit or rooted in tacit knowledge," this does not mean that all knowledge is tacit knowledge. It may say no more than does Campbell's remark, which we noted earlier, that qualitative, commonsense knowing is the building block and test of quantitative knowing. But it does mean that the idea of "explicit" is much more complicated than we (once) believed. Collins goes as far as to say: "There is nothing strange about things being done but not being told—it is normal life. What is strange is that anything *can* be told" (2010, 7).

Unlike Polanyi, who emphasizes the personal and individual aspects of tacit knowledge, for Collins the tacit is essentially social and is communicated by "hanging around" those who have it.[7] In speaking of *relational* tacit knowledge, his position is that "any one piece of relational tacit knowledge can be made explicit, because the reason it is not explicit is contingent on things that can be changed. But all relational tacit knowledge cannot . . . be made explicit at once" (98). To give an example of one kind of relational tacit knowledge, there may be circumstances where knowledge is *concealed*. In a study of the national introduction of electronic recording systems for children in need or at risk in England and Wales, there were occasions when some social workers who, when talking to service users, explained the reason for using the lengthy and elaborate forms as stemming from a professional obligation to complete a new administrative system (Shaw et al. 2009). This was true but concealed important parts of the truth. There are also instances of what Collins calls *unrecognized* knowledge, as when someone knows something but does not know what is the important part of that knowledge and so cannot tell someone. We can envisage such instances on the part of members of a community, as described above when a stranger enters that community. This and other kinds of relational tacit knowledge are tacit only in a loose sense because explicit knowledge is open to those in the in-group. This poses the question whether all relational tacit knowledge is explicable. Yes, Collins says, if the contingencies can be made to go away. But this is a very big "if"—if people stopped keeping secrets, if we knew what was in other people's heads, and so on. "In society as we know it there will always be secrets, mismatched saliences, and things that are unknown but may be about to become known" (Collins 2010, 98).[8]

Bicycle riding, thanks to Polanyi, has become a paradigm of tacit knowledge. We speak of some skills as being "just like riding a bike," meaning the body's "knowledge" of this skill once acquired is not forgotten or at least is easily retrieved. Collins argues that it is possible to make learning to balance explicit—but there are somatic limits to our ability to follow any explicit instructions. It is somatic-limit tacit knowledge—the *second* form of tacit knowledge. We know that whether or not people can use sets of instructions, "they can do many things better if they do not process the instructions with the conscious mind" (Collins 2010, 101). In broad summary, when first learning a skill of this kind we are more likely to give self-conscious attention to explicit rules, but experts mostly will exercise skills without paying conscious attention to the process. I recall in days long past that I employed a skilled typist to type my doctoral dissertation. I discovered that she had little or no idea what she had been typing. In other words, she did not need to have meaningful words passing through her head in order to type swiftly and efficiently. Her brain "knew" where the letters were without having to make sense of the words.

Collins employs a distinction from chess between foreseeing a sequence of moves in advance and seeing a pattern of play. Example 7.3 illustrates how another game metaphor may illuminate how social workers struggle to make sense of otherwise tacit knowledge of work that goes either well or badly.

EXAMPLE 7.3

"Game Plans" and Tacit Knowledge

In a study of social workers' accounts of work that had gone well and badly, a team leader said of his practice: "I have a kind of game plan in my head which I just do. . . . I think most social workers, or *good* social workers rather, know their capabilities, know their skills, know what they should be doing and they get on with it."

The possession of a *game plan* depended, in part, on "staying in the ball park." A probation officer described a presentence report in which a very low tariff sentence was recommended but where a custodial sentence was given. In attempting to reflect on the extent to which he felt he had misread the situation, the probation officer concluded, "If the judge had said, 'No. Conditional discharge isn't serious enough, I'll fine her,' then I'd say, 'Fair enough.' I was in the ball park, you know."

This practitioner believed he had been outside the ball park in the sense that he had unwittingly but completely misjudged the rules of the game. "I chose the timing wrongly . . . and the outcome was to damage our relationship for a couple of months. . . . It was that judgment, it was actually about timing, and I think your timing goes when you're under pressure."

Applying a general "game plan" to a specific situation was believed to require the possession of "good timing" on the part of the practitioner. A children's social worker attempted to make explicit why a case seemed to have gone well: "I felt the timing for the children was essential, that they were placed with the adopters when the baby was quite small. . . . I felt a sense of satisfaction that I had managed to set the plans up, not made too quick a decision in either direction, and that the outcome was good for everybody."

Whatever personal game plans may be in place, there remains an element of unpredictability near the heart of social work practice. According to those social workers, whether work goes well or badly is not the product only of rational and intelligent planning. At its simplest, it is reflected in the assumption of one social worker that "there's always an element of good fortune in these things."

Source: Shaw and Shaw (1997/1999).

Collins names the *third* form of tacit knowledge "collective tacit knowledge": "the irreducible heartland of the concept" (2010, 119). "The central domain of tacit knowledge—collective tacit knowledge—is beyond explication" (153). "We can describe the circumstances under which it is acquired, but we cannot describe or explain the mechanism nor build machines that can mimic it" (138). He alludes to how the "close studies of science . . .

show that there are no well-structured domains, no areas which are completely formalized and completely calculable." It is "the *appearance* of complete calculability that has to be explained" (146, emphasis added).

Take, for example, decisions about when to improvise in social work when applying organizational rules in practice. "It is only humans who have the ability to acquire cultural fluency . . . the ability to absorb ways of going on from the surrounding society without being able to articulate the rules in detail" (125). In response to the interest in social work on the roles animals may play, Collins would insist, "there are no groups of vegetarian dogs, arty dogs, nerdy dogs, dogs that believe in witches, and dogs that understand mortgages—they are all just dogs" (125).

For social workers there are actions that require different behaviors in different contexts and require different interpretations according to in which context they take place. Thus, what count as sensible questions and answers depends on context. When we talk about how an individual acquires knowledge, we may say it is through immersion in society—by participating in the talk and practices of society. To become fluent in a language "is to master the tacit knowledge inhering in the conceptual life of society" (135).

There is organizational knowledge on which members draw "without being aware that they are doing so and certainly without being able to express fully what they know as a result of their membership in the organization" (144). We may think of social work organizations in light of this, for example, of what happens when things stop running smoothly. In doing so intuition remains a useful term "so long as we remember that it is 'wisdom based on experience'; that is mysterious enough. This kind of 'intuition' can be gained through practice and socialization, including the acquisition of interactional expertise" (149).

Collins explores "the difference between the knowledge of experts when they make assessments of technical value and the knowledge of non-experts" (2014, 722). He describes a case study of being sent a "heterodox" physics paper on gravitational wave theory (his area of research within sociology), sending it to twelve senior physicists in the field, and assessing his own ability to use the criteria they employed to assess the technical level of the paper. He "aims to compare experts to non-experts when they are faced with heterodox claims" (729) and to show that "the tacit knowledge of the expert community is unavailable to non-experts and also to show what this means for the way judgements are made" (729). In all this he is not insisting that expert knowledge is unquestionably valid. "Scientists are neither Gods nor charlatans; they are merely experts" (Collins and Pinch 1988, 143).

Social Work Science and Judgment

Common sense, expertise, tacit knowledge—each occasions moments of judgment. What is involved when critical reflection, judgment, and discretion are exercised in social work science?

Judgment and discretion form part of all approaches to social work science. John Stuart Mill developed his canons of proof to deal with this, so as to avoid a need to refer to theory. This led to "a marriage of an empirical interpretation with the certainties of deductive logic," which "became known is the hypothetico-deductive model of scientific explanation" (Hughes 1980, 50). Empiricism, with its main distinguishing feature being relating the observable to the observable, is in effect "a system of trial and error" (57). Positivist science does not deny that people pursue goals or have interpretations and reasons. How, then, are they treated by positivists? Often through the idea of "role expectations" existing external to the person and thus producing social patterning. Motives are patterned by socialization.

There is a profound difficulty in how theory is handled in this position. Take, for example, interpreting associations, which are never perfect. Do they prove or disprove a theory? The way we interpret associations is usually post hoc—"All kinds of rationalisations, plausible and sometimes less than plausible, are entered into to make the associations theoretically 'interesting'" (55). This is because social science generalizations are "accidental," which does not mean they lack either interest or significance but that they are the consequence of which variables were thought important enough to be brought into a study, and then which choices among typically many associations have been tested.

Because motives are prior to and independent of behavior, this leaves the problem of measuring motives. Inferring back from behavior to motives is a constant hazard typically encountered through the problem of the relation of what people say to what they do. One piece of behavior can be part of many different actions.

There are examples where for all practical purposes one cannot infer motives simply from behavior—for example, suicides. Interpreting actions as governed by *rules* poses the same problem as motives, in that behavior may be consistent within a number of different rules, so rules cannot always be inferred from behavior. So to be rule governed is not the same as mere regularity. For example, it is possible to make a mistake in relation to rules, though not in relation to mere regularity. Our actions are made accountable and the world made understandable.

Induction, Abduction, and Deduction

Relating questions of judgment to science links us to the discussion of induction and deduction in chapter 1 and to the arguments for abduction. Logicians commonly had divided arguments into two subclasses: the class of deductive arguments (necessary inferences) and the class of inductive arguments (probable inferences). Charles Peirce began to hold that there were two distinct classes of probable inferences, which he referred to as inductive inferences and abductive inferences. For Peirce, induction in the most basic sense is argument from random sample to population. Peirce came to see the scientific method as involving three phases or stages: abduction (making conjectures or creating hypotheses), deduction (inferring what should be the case if the hypotheses are the case), and induction (the testing of hypotheses).

> Scientific method begins with abduction or hypothesis: because of some perhaps surprising or puzzling phenomenon, a conjecture or hypothesis is made about what actually is going on. . . . Scientific method then proceeds to the stage of deduction: by means of necessary inferences, conclusions are drawn from the provisionally-adopted hypothesis about the obtaining of phenomena other than the surprising one that originally gave rise to the hypothesis. Conclusions are reached, that is to say, about other phenomena that must obtain if the hypothesis should actually be true. These other phenomena must be such that experimental tests can be performed whose results tell us whether the further phenomena do obtain or do not obtain. Finally, scientific method proceeds to the stage of induction: experiments are actually carried out in order to test the provisionally-adopted hypothesis by ascertaining whether the deduced results do or do not obtain.[9]

More specifically,

> abduction became defined as inference to and provisional acceptance of an explanatory hypothesis for the purpose of testing it. Abduction is not always inference to the best explanation, but it is always inference to some explanation or at least to something that clarifies or makes routine some information that has previously been "surprising," in the sense that we would not have routinely expected it, given our then-current state of knowledge.

Sowa (2005) visualizes this process in figure 7.1, which shows an agent who repeatedly carries out the stages of induction, abduction, deduction, and action. The arrow of induction indicates the accumulation of patterns that have been useful in previous applications. The crystal at the top symbolizes the elegant but fragile theories that are constructed from chunks in the knowledge soup by abduction. The arrow above the crystal indicates the process of belief revision, which uses repeated abductions to modify the theories by expansion, contraction, revision, and analogy. At right is a prediction derived from a theory by deduction. That prediction leads to actions whose observable effects may confirm or refute the

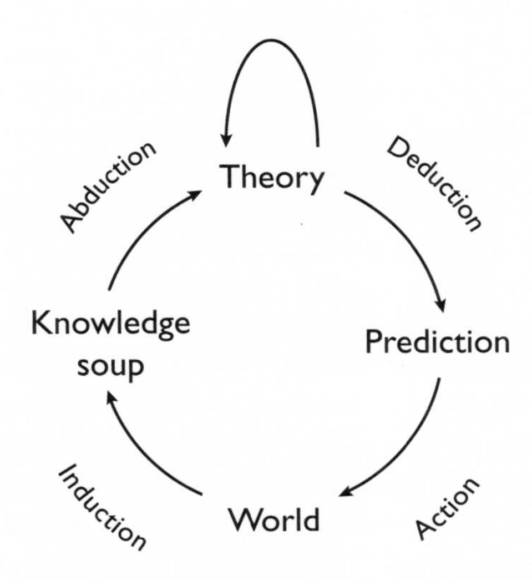

Figure 7.1 The Cycle of Induction, Abduction, Deduction, and Action

theory. Those observations are the basis for new inductions, and the cycle continues.

> Learning is the process of accumulating chunks of knowledge in the soup and organizing them into theories—collections of consistent beliefs that prove their value by making predictions that lead to successful actions. Learning by any agent—human, animal, or robot—involves a constant cycling from data to models to theories and back to a reinterpretation of the old data in terms of new models and theories. Beneath it all, there is a real world, which the entire community of inquirers learns to approximate through repeated cycles of induction, abduction, deduction, and action.
>
> (Sowa 2005)

Judgment and Taste

Understanding the role of judgment in social work science further can be sharpened by considering how it differs from matters of taste. "Objectivity is not something we can straightforwardly be against—how else do we commend our own stories but through some warrant bearing a family resemblance to objectivity?" (Shapin 2012, 170). The characteristic approach of those seeking to understand the social aspects of science has been to "take supposedly objective claims . . . and show their impurity, the presence in them of disparate, subjective things" (171), and thus to historicize objectivity, such that "our scientific knowledge is about the world and it is also irremediably about us, as knowers" (171). Subjectivity is called on to deflate objectivity, as a Trickster—"a disruptive and disordering influence on knowledge . . . making a mess out of everything" (171).

However, subjectivities have their modes and specifics. Take, for example, the idea of matters of taste, and whether some issues in social work science are questions of taste. Shapin cites the Latin tag *de gustibus non est disputandum*—"there's no accounting for taste," as we say—and in consequence there is nothing we can say to one another that will provide grounds to alter one's position. Shapin distinguishes what are matters of judgment and matters of taste. Matters of judgment are "discussable," and evidence and argument can be brought to bear. "Taste" is "incorrigibly subjective; it is private; and for that reason there is nothing that can be said about it that has any consequence" (172). If I say that a Vermeer painting—*The Little Street*, for instance—is beautiful, I am saying something about me but also something about the painting. I do not have to assert universal aesthetic standards, but still "I mean it to be understood that there is something about the painting that is the occasion of my saying so and ought to cause you to respond as I respond" (175).

How is this of interest to social work? There has been too little interest in questions such as: How do we learn what a good social worker is? How does one become an enthusiast for, say, randomized control trials? How does one become a good researcher? There has been an unduly relativist tendency in social work to see choices in some areas as matters of taste or preference, though to say that something is a matter of taste is itself a judgment—that is, that two tastes are equally good.

Scientific rules are not unlike rules in art: they are indeterminate and act as guides in the art of making scientific discoveries and in assessing the value of a scientific claim. Roald Hoffmann, a theoretical chemist, remarks:

> I think poetry and a lot of science—theory building, the synthesis of molecules—are creation. They're acts of creation that are accomplished with craftsmanship, with an intensity, a concentration, a detachment, an economy of statement. All of these qualities matter in science and art. There's an aesthetic at work, there is a search for understanding. There is a valuation of complexity and simplicity, of symmetry and asymmetry. There is an act of communication, of speaking to others.[10]
>
> (Wolpert and Richards 1997, 23)

Interpretations are not self-evidencing. Theorizing in social work science is difficult and requires "the need for theory to stay rather closer to the ground than in the mainstream sciences" (Geertz 1973, 24). The essential task of theory building "is not to codify abstract regularities but to make thick descriptions possible, not to generalize across cases but to generalize within them" (26). He suggests this is exactly like the process of clinical inference. "The aim is to draw large conclusions from small, but very densely textured, facts" (28). In doing so he insists on the need "to try to resist subjectivism on the one hand and cabbalism on the other" (29–30). On the former, "nothing will discredit a semiotic approach to culture more quickly than allowing it to drift into a combination of intuitionism and alchemy, no matter how elegantly the intuitions are expressed or how modern the alchemy is made to look" (30). On the latter, he warns of "the danger that cultural analysis . . . will lose touch with the hard surfaces of life" (30).

Being Accountable

Returning to the point about making actions accountable, Pithouse and Atkinson (1988) explored the character of "case talk." In the case text presented by these authors, causal attributions are present and are part of the fabric of the narrative. They report a social work team meeting in which a social worker and team leader discuss a "case" of a girl on a supervision order who had been shoplifting again. In trying to explain this behavior the social worker says, "But her criminality is in [*pause*] er, strange really, it's almost a mania. It has a quality about it that's almost psychologically driven. I don't know if that's the proper use of the term 'psychology' but—you know—the drive is there, er, because of an abnormal psychology, there's something there all right" (Pithouse and Atkinson 1988, 189).

A moment later, this is echoed by the team leader confirming, "yes, there's something *there* in the family." The social worker suggests a determinist view of the family's behavior as "psychologically driven." Moreover, this kind of causal explanation appears to carry with it a message about the diminished performance expectations that can reasonably be anticipated from the social worker.

A similar and explicit example of a causal case account being used as a professional safety net occurs later in the same text. In drawing the threads together into the final denouement of the narrative, the social worker asserts an argument about culture: "This is quite a delinquent family, we're not going to change that fact of their life-style—it's cultural—I'm not going to change this you know—I don't intend to" (192).

Whether in any precise sense cultural or psychological factors are at work is largely immaterial to whether or not the account is accepted and "honored" by the audience. A causal account is presented to justify and gain colleague and management agreement for an expectation of *limited* and *imprecise* effects from the social worker's intervention and for a style of work described earlier by the social worker as "continually chipping away at the family" (191).

In exploring tentative hypotheses for "trigger" events of causal accounts, it may prove of central importance to understand how social workers explain and account for things "going wrong." While the very idea of "making a mistake" assumes rule-governed, rather than determined, behavior, it is arguable that we often refer to *causes* when something has gone wrong (Hughes 1980, 83, 89). Within the category of "going wrong" we may include "mistakes," "crises," a worsening of the client's "problem" notwithstanding the social worker's efforts, and also when positive or even negative predictions or explanations appear to be contradicted by events. Thus, negative predictions may come unstuck, and, although this may be good for the client, it is to us still an example of "going wrong."

Lyman and Scott's (1970) analysis is helpful at this point. They distinguished two kinds of account, "excuses" and "justifications." An excuse is offered when someone accepts that the action is bad, wrong, or inappropriate but responsibility is denied. Justifications take the form of accepting responsibility but denying the pejorative quality of the action. Taking this as the essence of the distinction, accounts can thus be more fully categorized within figure 7.2.

Because action includes motive, any action-description is defeasible—it is always possible in principle to argue against it. House (1980), the evaluation theorist, developed one significant implication of this by seeing evaluation as an act of persuasion. He argues that the logic of evaluation is not so much rational evidence as it is persuasion and argumentation. Evaluations never yield certain knowledge. "Subjected to serious scrutiny, evaluations

	Responsibility accepted	Responsibility denied
Blame accepted	Confession	Excuse
Blame denied	Justification	Repudiation

Figure 7.2 Social Work "Accounts"

always appear equivocal." The aspiration after certain knowledge "results from confusing rationality with logic. They are not identical. . . . Evaluations can be no more than acts of persuasion" (72). They are acts of argumentation, not demonstration. This is as true for quantitative methodology as it is for qualitative, despite the fact that "statistical metaphors . . . give a semblance of certainty and unequivocality to evidence" (74). In summary, "evaluation persuades rather than convinces, argues rather than demonstrates, is credible rather than certain, is variably accepted rather than compelling" (73). House characteristically interleaves considerations of logic and moral positions. He repeatedly claims that normative considerations have been neglected and that "if this is a weakness in the conduct of science, in evaluation it is a fatal flaw" (251).

The emphasis given in this chapter to understanding is likely to seem disproportionate when compared with mainstream social work science in the United States. Yet it chimes with my stress on a social work imagination in the introduction and, with a minor switch of terms, with the Chicago sociologist Everett Hughes, when he commended "the intensive, penetrating look with an imagination as lively and as sociological as it can be made" (1984, xix). Yet this is not all that should be said. What passes for social reality stands in immediate relation to the distribution of power. Giddens warns against "a concern with 'meaning' to the exclusion of the practical involvements of human life in material activity . . . a tendency to seek to explain all human conduct in terms of motivating ideals at the expense of the causal conditions of action; and a failure to examine social norms in relation to asymmetries of power and divisions of interest in society" (1993, 163–164). It is to the implications of this for social work science that we turn in the following chapter.

Taking It Further

Reading

Geertz, C. 1983. "Common Sense as a Cultural System." In *Local Knowledge*. New York: Basic Books.

Williams, G., and J. Popay. 1994. "Lay Knowledge and the Privilege of Experience." In *Challenging Medicine*, ed. J. Gabe et al., 118–139. London: Routledge.

There is no shortcut to getting a satisfying grip of the range of questions in this chapter. For straightforward starters, both Geertz and Williams and Popay are different but excellent entry points.

Collins, H. 2010. *Tacit and Expert Knowledge*. Chicago: University of Chicago Press.
Giddens, A. 1993. *New Rules of Sociological Method*. Stanford, Calif.: Stanford University Press.
Schutz, A. 1967. "Common-Sense and Scientific Interpretation of Human Action." In *Collected Papers*, vol. 1: *The Problem of Social Reality*. The Hague: Martinus Nijhoff.

I have tried to tackle the "big" questions of tacit knowledge, expertise, common sense, and how these are inextricably tied in to science. Having read Collins, Giddens, and Schutz, one will be chastened but more judicious.

Task: Tacit and Explicit Knowledge in Social Work

Consider an example related to social work science. It may be:

- Doing social work practice with a given case, based on a familiar model of intervention
- Using a standardized social work assessment instrument
- Preparing a report for a court or a case conference

Distinguish the relational, somatic, and collective tacit knowledge and see how far knowledge can be rendered explicit and be "mimeographed" (or, to use a more familiar social work term, "manualized"). Recall in doing so what Collins means when we say something has been made explicit and what we intend by saying "can" or "cannot" be rendered explicit.

[8]

Social Work Science and Justice

In mapping the range of scientific allegiances to social justice, I draw on the evaluation field to distinguish nonpartisan, multipartisan, reformist, and critical theory positions. I then go on to ask how positions on social justice are related to methodological choices, consider ways in which we should think of citizens and service users as having scientific knowledge about social work that shapes the justice agenda, and conclude the chapter with an exploration of arguments about insider and outsider status, standpoint theory, and feminist social work science.

Public intellectuals "have been giving up the old ideal of the public relevance of knowledge." So wrote C. Wright Mills in 1955 (Mills 1963, 599). "Among the men of knowledge, there is little or no opposition to the divorce of knowledge from power, of sensibilities from men of power, no opposition to the divorce of mind from reality . . . and when men of knowledge do come to a point of contact with the circles of powerful men, they come not as peers but as hired men" (604–605). He laments "the malign ascendancy of the expert, not only as a fact but as a defense against public discourse and debate" (608). "Knowledge is no longer widely felt as an ideal; it is seen as an instrument" (606), such that "knowledge does not now have democratic relevance in America" (613).

Justice is the third of the leitmotifs within all social work science. The waters are muddy. For example, American social work often acclaims the Milford Conference for its "success" in formulating an "orderly outline" for the social work profession (Walter West in the foreword to Milford Conference 1929).[1] Yet in ordering social case work as the heart of the profession, the conference report tells us that "social case work deals with the human being whose capacity to organize his own normal social activities may be impaired by one or more deviations from accepted standards of normal social life" (Milford Conference 1929, 16). Among these deviations we find, to use the terminology of the conference's lengthy list, bad housing, common law marriage, dependent old age, foreignness, homelessness, physical ill-health, lack of skill in trade, and seasonal employment. While the report acknowledges "there is not necessarily failure in self-maintenance when an individual is, for example, unemployed or financially dependent" (16), their characterization as deviations from the norm illuminates the model of society held by advocates of what then was seen as cutting-edge thinking in social work.

In order to find ways through these muddy waters and gain some perspective, I will step sideways from social work and set out a framework drawn from positions developed within the evaluation field as a way of distinguishing possible stances on the relationship of social work science to concerns for individual and social justice. Four possible positions are set out in table 8.1.

TABLE 8.1
Possibilities for allegiance in social work science

ALLEGIANCES	EVALUATION IDENTIFIER
Nonpartisan	Michael Scriven
Multipartisan	Lee Cronbach
Democratic reformism	Ernest House
Critical science	Thomas Popkewitz

Evaluation and Scientific Allegiances

Nonpartisan Evaluation and Science

An argument for the social work scientist to stand aside from any allegiance should not be regarded as synonymous with a simple acceptance of the objectivity of scientific knowledge. At the core, for example, of Michael Scriven's argument is the view that evaluation assigns merit or worth. "The fact that evaluation, as currently conceived by most of its practitioners, can only lead to descriptive conclusions, not to evaluative ones," represents in his view "the terminal disease" of evaluation (1986, 10). However, he makes careful distinctions between judgments of the merit of what is evaluated (the "evaluand"), *explanations* (through formative evaluation), and *recommendations* ("remediations"). "Bad is bad and good is good, and it is the job of evaluators to decide which is which. And there are many occasions when they should say which is which whether or not they have explanations or remediations" (19).

"Goal-free evaluation" is Scriven's ideal means of reaching an evaluative judgment, by which he means that evaluators should proceed as if under a veil of ignorance regarding the goals held by program managers and other stakeholders. Goals bias the evaluator. Rather than ask, "there are the objectives. Have they been achieved?" Scriven would prefer, "Here is the program. What are its effects?" This leads to an evaluation practice in which the lead evaluator is almost inevitably an outsider, and for the outside evaluator "both distancing and objectivity remain correct and frequently achievable ideals" (1997, 483). A benchmark assessment of the evaluation field concluded that "his work on valuing is so complex, subtle and so full of information, we learn new things each time we read it" (Shadish et al. 1990, 118).

Multipartisan Evaluation and Science

Should the very notion of being partisan on behalf of quite contrary positions sound like equivocation, I want to refer again to the work of Lee Cronbach here and in the closing chapter. While I demur from him at several points, Cronbach is one of the most important and least appreciated evaluation theorists. His work is almost never cited in social work science or research, yet the work of this "tough-minded master of conceptual distinctions" (Scriven 1986, 15) is "a brilliant *tour de force*, unusually rewarding if closely read, trenchant in analysis of the *status quo*, and creating truly unique alternatives sensitive to the scholarly need for general knowledge and the practitioner need for local application" (Shadish et al. 1990, 375).

His pluralist view of methodology echoes his conception of the policy context in which evaluation is located. With characteristic dry wit he remarks: "The very proposal to evaluate has political impact. To ask about the virtue of Caesar's wife is to suggest she is not above suspicion" (163). For Cronbach and his colleagues, a theory of evaluation "must be as much a theory of political interaction as it is a theory of how knowledge is constructed" (52–53), flowing from the recognition that "politics and science are both integral aspects of evaluation" (35). So the argument for a social theory is more like an argument for a political case than a natural science explanation (Cronbach 1986). "Time and again political passion has been a driving spirit behind the call for rational analysis. . . . The very call for value free inquiry was politically motivated, generally by discontent" (Cronbach et al. 1980, 35). Hence he regards the role of the evaluator as a multipartisan advocator. "It is too much to ask for twenty-twenty foresight, but it is the role of evaluation to supply spectacles for every nose, to bring as much into focus as possible" (155). The evaluator "can be an arm of those in power, but he loses most of his value in that role if he does not think independently and critically. He can put himself in the service of some partisan interest outside the centre of power, but there again his unique contribution is a critical, scholarly habit of mind" (67).[2]

While Cronbach would have stood in a quite different place from C. Wright Mills, perhaps he would have concurred with the latter, who said:

> As a type of social man, the intellectual does not have any one political direction, but the work of any man of knowledge, if he is the genuine article, does have a distinct kind of political relevance: his politics, in the first instance, are the politics of truth, for his job is the maintenance of an adequate definition of reality. In so far as he is politically adroit, the main tenet of his politics is to find out as much of the truth as he can, and to tell it to the right people, at the right time, and in the right way. . . . The intellectual ought to be the moral conscience of his society, at least with reference to the value of truth, for in the defining instance that *is* his

politics. And he ought also to be a man absorbed in the attempt to know what is real and what is unreal.

(Mills 1963, 611)

Example 8.1 illustrates how Cronbach saw evaluation within a context of political accommodation as both conservative yet committed to change.

EXAMPLE 8.1
Cronbach on Active Gradualism

Evaluation's "role is not to produce authoritative truths but to clarify, to document, to raise new questions, to create new perceptions" (Cronbach et al. 1980, 53). "This is the stance of a friendly critic, not of a person who sees the system as either beyond reproach or beyond repair" (157). His position is to see evaluation as a "handmaiden to gradualism" (158)—"evaluation is both conservative and committed to change" (157). "To be meliorist is the evaluator's calling. Rarely or never will evaluative work bring a 180-degree turn in social thought. Evaluation assists in piecemeal adaptations: perhaps it does tend to keep the status very nearly quo" (157).

Cronbach was not a policy science optimist. "In debates over controversial programs, liars figure, and figures often lie; the evaluator has a responsibility to protect his clients from both types of deception" (38), yet "we shall not advise the evaluator to avoid opportunities to be helpful; he is hired to improve public services, not to referee a basketball game" (18). He is "to act as a buffer between observers and political actors . . . not to censor their communications but to ripen them" (71).

"The public should continue to hear many voices carrying discordant messages, but it would benefit if more speakers were disciplined by collegial discourse" (71). "The larger the role of experts in governance, the more difficult it becomes for ordinary citizens to give direction to action. . . . Insofar as information is a source of power, evaluations carried out to inform a policy maker have a disenfranchising effect. . . . An open society becomes a closed society when only the officials know what is going on" (95). There is a historic tension between elitism and participation. "The rationalist ideal of efficiency is in tension with the ideal of democratic participation. Rationalism is dangerously close to totalitarianism" (95). Information that is correct and comprehensive is of no use if it is not credible and comprehensible. Thus the evaluator "is not to see himself as a philosopher-king," and "we advise him to respect the citizen's right to decide" (72). In ways not wholly unlike Weber, when speaking of the evaluator as public scientist Cronbach and his colleagues suggest a set of personal values:

1. The evaluator should "not attempt to evaluate a program with whose basic aims he is not in sympathy" (208). In words that anticipate later writing on appreciative evaluation, they think that oversympathy is much less a problem, so long as there is a commitment to effective programs.
2. Openness to good and bad news in data collection and impartiality in interpretation. This is not the same as value neutrality but is a consideration of "the facts from the relevant, no doubt conflicting, perspectives" (209).
3. "Having done his professional job, he puts off his professional robes and, if he chooses, speaks up for what he as a citizen favours" (209).

Reformist Evaluation and Science

The exemplar I draw on in this regard is Ernest House. The bridge—and distinction—from Cronbach's work can be seen in Cronbach's expression of House's position:

> House would have the evaluator seek out the data most likely to support programs that benefit the dispossessed—not in a spirit of misrepresentation but in a spirit of highlighting what the well-to-do overlook. Moreover, House would have the investigator interpret the facts from the standpoint of citizens who are least well-off and would have him try to persuade the entire [policy-shaping community] to do what will best serve that subgroup. The facts reported would be legitimate, honest fruits of research; the partisanship would lie in the emphasis given them.
>
> (Cronbach et al. 1980, 209)

House is not happy with many existing expressions of scientific allegiance on three counts. First, their value positions are subjectivist or descriptive. Whether they are directed at managers, professional elites, or consumers, they distance evaluation from prescriptive recommendations. Second, their epistemology is either objectivist or intuitionist. Third, the political assumptions are uniformly pluralist. While he does not level a charge of political complacency, he complains that they are all based on essentially Enlightenment philosophies. "All assume that increased knowledge will make people happy or better satisfied in some way" (House 1980, 64).

Extensive discussion of House's response is captured in his major book, *Evaluating with Validity* (1980). With respect to justice his position is twofold. First, he argues that the logic of evaluation is not so much rational evidence but persuasion and argumentation. Evaluations never yield certain knowledge. "Subjected to serious scrutiny, evaluations always appear equivocal." The aspiration after certain knowledge "results from confusing rationality with logic. They are not identical. . . . Evaluations can be no more than acts of persuasion" (72). They are acts of argumentation, not demonstration. This is as true for quantitative methodology as it is for qualitative, despite the fact that "statistical metaphors . . . give a semblance of certainty and unequivocality to evidence" (74). In summary: "Evaluation persuades rather than convinces, argues rather than demonstrates, is credible rather than certain, is variably accepted rather than compelling" (73). House characteristically interleaves considerations of logic and moral positions. He repeatedly claims that normative considerations have been neglected and that "if this is a weakness in the conduct of science, in evaluation it is a fatal flaw" (251). He developed the implications of this more explicitly in the idea of deliberative evaluation (House and Howe 1999).

Second, House has developed extensive arguments for a reformist, social justice purpose for evaluation. He is not indifferent to the aesthetics of evaluation. Yet for him, "truth is more important than beauty. And justice more important than either" (117). Utilitarian theories of justice, based on the greatest net satisfaction, provide clear criteria for evaluating yet tend to favor the upper classes, lead to judgments that "do not square with one's moral sensibilities" (134), and often lead to oversimplification. Pluralist and intuitionist theory comes closest to common sense and everyday judgments of justice, and it fosters a valuable emphasis on portraying the opinions of stakeholders. Yet "the threat of relativism" (134) jeopardizes consistency of application, and it tends to place too high a value on professional judgment. "In essence, the pluralist model confuses issues of interests with conflicts of power. It can balance only those interests that are represented—typically those of the powerful" (House 1991, 240).

House's conception of justice finds its most practical application in his detailed arguments for a fair evaluation agreement (1980, chap. 8) and in his broader arguments for evaluation ethics (1990, 1993). His replacement of traditional ideas of objectivity with impartiality undergirds his explicit reformist stance. He advocates a moral basis of evaluation resting in principles of moral equality, moral autonomy, impartiality, and reciprocity, without being sure how they are to be balanced against one another in every situation. None should have particular priority, and decisions should be made in pluralist fashion, with considerations of efficiency playing a part but with justice as prior. In other words, he offers a strong dose of John Rawls, and also of more intuitionist approaches, with a slight dash of utilitarian efficiency. He develops the example of negotiating a fair and demanding evaluation agreement (1980, chap. 8), in which all participants should meet the demanding conditions that they

- Not be coerced
- Be able to argue their position
- Accept the terms under which the agreement is reached
- Negotiate—and this is not simply "a coincidence among individual choices" (1980, 165)
- Not pay excessive attention to one's own interests
- Adopt an agreement that affects all equally
- Select a policy for evaluation that is in the interests of the group to which it applies
- Have equal and full information on relevant facts
- Avoid undue risk to participants arising from incompetent and arbitrary evaluations

House defends this reformist position. To critics who say he is biased to the disadvantaged, he responds: "It seems to me that making certain the interests of the disadvantaged are represented and seriously considered is not being biased, though it is certainly more egalitarian than most current practice" (1991, 241–242).

A limitation of House's position is that he tends to assume an overly consensual view of society, which limits his reformist position. He believes that although philosophers, legislators, and the public disagree, they do so "within the frameworks of overall agreement about fundamental democratic values" (1995, 44). Yet his thinking on justice is subtle, whether he warns that evaluation practice is not *determined* by our views of justice or distinguishes cultural and moral relativity in accepting the proper diversity of concepts of justice.

Critical Evaluation and Science

Critical evaluation is "critical" in that evaluation problems are conceptualized within their social, political, and cultural context and reflect the approach of critical theorists whose work we referred to in chapter 1 and elsewhere. It is "critical" evaluation, to borrow a phrase from Popkewitz's essay on the critical theory paradigm, because it "gives reference to a systematic inquiry that focuses upon the contradictions of . . . practice" (1990, 46). The critical theorist reconstructs the rules of evaluation to reflect the understanding that the language categories of research and evaluation are historically related to larger social and moral issues of production and reproduction. "Methodology is concerned with the moral order . . . presupposed in the practice of science. It is the study of what is defined as legitimate knowledge" (51–52). It entails a theoretically based analysis marked by skepticism toward the forces of social regulation and distribution and sometimes also to empirical evidence. "Whatever it is must not be taken at face value" (Harvey 1990, 8). Lather refuses to demonize other persuasions and insists that "there are no innocent positions" (1991, 85). Critical evaluators will reject the idea of science as cumulative and progressive development. Evaluation work is more fragmented at this point, and it will help to shift to more general discussions of science and social justice.

Methodology and Justice

We noted Giddens's observation, quoted at the end of chapter 7, regarding the neglect of power in hermeneutic understandings of science. Where might we stand on this question? More generally, how should we regard the relationship between methodology and social work science?

Qualitative research, so Packer insists, is not a set of techniques but a "basis for a radical reconceptualization of the social sciences as forms of inquiry in which we work to transform our forms of life" (2011, 3). Gergen also counters to some extent when he says that most discourse analysts "use research to further ends they value. They may . . . use research as *social critique*. For example, they will point out ways in which academic discourse excludes the less educated from understanding" (2009, 65). However, is it legitimate to ask: What if we suspect participants misunderstand their form of life? "This is the troubling suggestion made most powerfully by Karl Marx" (Packer 2011, 271), who added the place of "critique" with the aim of "emancipation." Critical science stems from Marx, Hegel, and Kant, then through the Frankfurt School and thence to Habermas.

Marx had much to say about alienation—the process whereby workers are separated from one another, from the products of their labor, and from the activity of work itself. Alienation exerts power such that workers are unaware they are being exploited, thus producing false consciousness. Down through subsequent theorizing the consequent vision for research has been an emancipatory one. Packer's conclusion to the "what if" question is that we "still need to take their understanding into account. We do not need to accept the understanding that participants display in an interaction, and our analysis does not need to stop there. But it does need to start there. . . . We cannot critique participants' understanding unless we first figure out what it is" (2011, 267). For some writers this includes a more general skepticism about methodology of any kind.

The Frankfurt School, including Horkheimer, Adorno, Marcuse, Erich Fromm, and Walter Benjamin, suggested that Marx had failed to give adequate place to cultural factors in historical change. They dispersed to California, Switzerland, and New York with the rise of Nazism and hence had wider influence. They did not agree that there is logic to history—for them there was "no inevitable progress in history, no inevitable and perfect goal, no end to partiality. The path of history depends on people's ongoing efforts" (Packer 2011, 285). Critical inquiry does not have totally objective analysis, and "scientific rationality should never be completely trusted" (285). There was at best a caution about scientific method. Horkheimer, speaking during the 1930s early in the Frankfurt School, will stand for central emphases of critical science. People

> of good will want to draw conclusions for political action from the critical theory. Yet there is no fixed method for doing this: the only universal prescription is that one must have insight into one's own responsibility. Thoughtless and dogmatic application of the critical theory to practice in changed historical circumstances can only accelerate the very process which the theory aimed at denouncing.
>
> (Horkheimer 2002, v)

His stress is on how the current social situation influences scientific structures and not "sheer logic alone" (195). Ideas of the purpose and goals of research play their part, but these are not self-explanatory nor a matter of insight. It is not only "an intrascienfic process but a social one as well" (196). The purpose of critical thinking is not the better functioning of society. "On the contrary, it is suspicious of the very categories of better, useful, appropriate, productive, and valuable, as these are understood in the present order" (207). "In genuine critical thought explanation signifies not only a logical process but a concrete historical one as well. In the course of it both the social structure as a whole and the relation of the theoretician to society are altered, that is both the subject and the role of thought are changed" (211).

Packer observes that there is a recurring strand of pessimism in critical theory science, in terms we encountered in chapter 3. For example, Horkheimer opines that despite the proletariat's awareness of contradiction, even that is "no guarantee of correct knowledge" (213). Indeed, one can find oneself in opposition to views held by the proletariat. "It is the task of the critical theoretician to reduce the tension between his own insight and oppressed humanity in whose service he thinks" (221). In words apposite for some social work enthusiasm for critical theory he insists that it is to fall short of what is needed when "the intellectual is satisfied to proclaim with reverent admiration the creative strength of the proletariat and finds satisfaction in . . . canonizing it" (214). There is an "ever present possibility of tension between the theoretician and the class which he is thinking to serve" (215).

A hard-hitting critical position of this kind is relatively rare within social work. The remarks from a senior British social work academic in example 8.2 capture what perhaps is a more frequent position. His comments follow from generally identifying himself with the British Labour Party.

EXAMPLE 8.2

Politics and Critical Social Work Science

In social work and in the critical social sciences as a whole I come across a lot of people who talk a radical talk and see themselves as very much on the left, but they aren't politically active—aren't involved in any local or national political organization but channel their supposed radicalism solely into academic work. . . . I prefer the idea of mundane pragmatic political involvement to try and improve a few things in small ways. The same would go for social work research. I think the rhetoric of radicalism has its place but is usually less effective than getting your hands dirty—doing research commissioned by government, for example, commissioned evaluations and so on. There's a common position here, I think, of pragmatic ameliorative politics.

Source: Personal communication.

We began this part of the discussion by asking: What if we suspect participants misunderstand their form of life? Expressed in terms that make the implications for how one can argue a case for justice-based science more obvious, Kemp says, "the question is whether social scientists can ever justifiably claim to understand actors' interests better than those actors do themselves" (2012, 664). The question of how interests influence social work science is central to issues about constructivist positions, for example in the case of MacKenzie's arguments about the history of statistics that we touched on in chapter 3. Woolgar suggests that defenses of the position that actors' perceptions of their interests cannot be challenged have been made on grounds such as the incorrigibility of subjective preferences, the nonobjective character of normative judgments, or the centrality of actors' understandings in constituting the world (Woolgar 1981). Those who believe an objective evaluation of actors' judgments can be made have been criticized, for example, for imposing their values on others or for the outdated notion of objectivity. On the other hand, those who think actors' judgments are incorrigible have been accused of conservatism. Kemp wishes to maintain an argument that social scientists can make justified evaluations of actors' accounts, but without invoking objective real conditions. Hence he does not want to say we can be sure what people's "real interests" are. He argues "that it is possible for social scientific accounts to be preferable to lay accounts, but that this has to be justified on a case by case basis, through a dialogue with lay actors" (2012, 665).

Kemp discusses the work of the political and social theorist Steven Lukes. He speaks of power as domination, which will be present wherever it furthers, or does not harm, the interests of the powerful and bears negatively upon the interests of those subject to it. Lukes, so Kemp continues, accepts there may be differences, interactions, and conflicts among one's own interests yet maintains that one can distinguish between presently held preferences and real interests. This is because it is possible for social scientists to take an "external standpoint" in their study of the beliefs and activities of actors. This clearly echoes the critical theory position that it is possible to judge that someone has mistaken views about something.

I will cover Kemp's argument in relative detail because I believe it gives us helpful purchase on debates about empowerment and user-led research and science. Kemp is unhappy with retaining a hierarchy of interests, such that the scientist can trump the views of other actors, yet he rejects the position that actors' judgments are incorrigible. On the question of whether actors' interests are unitary—whether there may be conflicts among an agent's interests—Kemp's response is to distinguish between theories and interests. "We need to be sure that we are not taking the existence of multiple and conflicting theories as a sign that there must be multiple and conflicting interests," though he is not "against the *possibility* that agents may have multiple and conflicting interests" (671). He is here taking an antirelativist position—one can have conflicting theories, but this does not necessarily mean conflicting truths.

This poses the problem of how social scientists can tell whether their interest attributions are justified in cases where these clash with the accounts of agents. The nonjudgmental position says that where there is a clash between social scientists' and actors' views of their interests, we should always accept the latter. Contrary to this, Kemp wishes to suggest that it is possible, in certain circumstances, for social scientists to justify the preferable character of their accounts of interests. Social scientists "should attempt to demonstrate the value of social scientific understandings by identifying difficulties and problems that lay actors have been experiencing as a result of acting on their own problematic understandings, and showing how social scientific understandings can resolve these" (673). In the extended quotation in example 8.3 the various aspects of his argument are brought together.

Without unpacking the detailed implications of this argument, it has fairly obvious applications to how social work researchers can engage with service users, practitioners, policy makers, and the like, in cases where social work science seems to be at odds with others' understandings of

EXAMPLE 8.3
Kemp's Conclusions on Actors' Interests

In the first place, I am defending the value of an evaluative conception of interests. . . . Second, I am supporting the idea that agents' own evaluations of their interests need not be the last word on the matter; that social scientists are able to evaluate the interests of agents and can, in certain circumstances, justifiably disagree about what is most beneficial for agents. . . . The third aspect of my approach . . . is the contention that the interest accounts of both social scientists and actors are fallible, theoretically mediated. . . . The fourth aspect of my approach . . . does suggest that processes of justification must involve social scientists attempting to show how they can identify problems and limitations in actors' understandings of their interests, and trying to show how social scientific accounts can resolve these. . . . It demands that claims to be compelling be demonstrated in each particular case by social scientists managing to convince actors that the social scientific account will resolve problems that they are experiencing. This approach therefore incorporates a dialogic element. . . . Social scientific interest accounts are only legitimated if lay actors became convinced through processes of debate that such social scientific accounts are right.

Source: Kemp (2012, 676–677).

their interests. It has connections with reformist positions on the justice-led allegiances of social work science that we saw earlier in this chapter and, as Kemp himself acknowledges, has similarity with pragmatic positions, albeit Kemp would go further by bringing in a general criterion of public good.

Quantitative Science and Justice

It is sometimes suggested that quantitative methodology gives less weight to matters of justice. Gergen quotes how "statistics are human beings with the tears wiped off" and that they "often function to silence public opposition and obscure understanding" (60).[3] Gergen's argument is not without some force. There are reasons, tying back in to earlier discussions at the start of this book, why those committed to a position that sharply distances scientific knowledge from the various kinds of citizen knowledge are less likely to give weight to science having a direct (rather than mediated) role in fostering social justice. Thus even a sympathetic commentator such as Oakley notes that randomized control trials "tend not to take seriously the views of the participants." In most of the studies "the thousands of people who took part in them appear mainly as numbers, rather than as individuals with their own stories" (1998b, 108).

But there *are* advocates of randomized control trials who do not regard them as inherently at odds with a commitment to challenging inequality and disadvantage. Indeed, Oakley offers one of the clearest such examples. She says, speaking of her career, that

> I discovered that in our excitement to dismantle patriarchy I and other feminist social scientists had mistakenly thrown at least part of the baby out with the bathwater. Women and other minority groups, above all, need "quantitative" research, because without this it is difficult to distinguish between personal experience and collective expression. Only large-scale comparative research can determine to what extent the situations of men and women are structurally differentiated.
>
> (Oakley 1999, 251)

More recently she recalls how she came to understand that the alleged "crisis of epistemology" of Western culture "is simply a trick of the mind invented by theorists who've got nothing better to do. Reality does exist and so does the real stress and pain that derive from a completely non-random (unfair) distribution of life chances" (2014, 258).[4]

Oakley focuses on the early 1960s through to the early 1980s in the United States, when the enthusiasm for experimentation was dominated by "a clear-cut model of the relationship between public policy and the knowledge to be derived from well-designed and well-evaluated social experimentation" (1998b, 93)—a link that now is "widely conceived to be tenuous or non-existent" (93). In the period from the 1960s to the 1980s she suggests the most important factor was the U.S. government's "War on Poverty"/Great Society initiative "and a government decision in the form of a Presidential Executive Order mandating the evaluation of social programmes" (95). There was a huge jump in federal spending on evaluation in that period, such that rather than evaluation being retrospective it went hand in hand with new policy.

Time and place caution against assuming an overly easy transferability of thinking about the relation of methods and justice. In terms of *time*, the relation of quantitative science to broadly justice issues is illuminated by traveling back to an earlier understanding of the roles of experimental designs. Ernest Greenwood at Columbia and F. Stuart Chapin at Minnesota pioneered the application of experimental methods to social problems. Chapin first wrote on it in 1917 and authored a book on *Experimental Designs in Sociological Research* in 1947. Greenwood wrote *Experimental Sociology* in 1945. Greenwood's work was "stimulated by the social reform concerns of the Depression, and informed by a desire to establish the most effective methods of improving people's lives" (Oakley 1998a, 1240). Chapin and Greenwood both wrote about sociology in ways that spoke to an applied agenda, for example through experimental designs (Chapin 1936, Greenwood 1945)—a link between reform and experiment that was also apparent in Campbell's 1969 paper "Reforms as Experiments."

With regard to *place*, survey models must also be viewed through the prism of Western pluralism. It is certainly true that beliefs about society held by English-speaking heirs of a broadly Judeo-Christian civilization may not be shared by others. Harré (1989, 23) suggests that members of such societies believe that:

1. They are autonomous individuals.
2. Despite being trapped in a web of conventions and apparently inexorable natural order, they are agents.
3. They have both individually and collectively a past and a future and so have histories.

It is against this cultural backdrop that we must place the argument sometimes heard that survey research offers one of the most democratic forms of research method. Methods and procedures can be made visible and accessible to nonspecialists and specialists working in other disciplines, which makes for a research method mode more transparent and accountable than many other methods used by researchers (Hakim 1987, 48–49).

Regardless of whether such assumptions are actually true for all, some, or any members of Western societies, the egalitarian premises of survey research are sometimes rejected at all levels of society in developing countries. For instance, simple random sampling carries the implicit assumption that the views of an Arab sheik and a Moslem village woman will be of potentially equal significance in reaching persuasive sociological understandings. This assumption will often be challenged in Arab society both by those in positions of power and those of lower status and influence. Further, respondents may not share with Western individualism the assumptions about the circumstances in which "truth telling" ought to be expected.

Related to this, "conventional" (in Western eyes) methods of data analysis are assumed by European and American social scientists to be relatively neutral and value free. Survey researchers, it is presupposed, will almost routinely uncover evidence of variance and will produce tables, typically in the form of cross-tabulations, which will differentiate replies according to this or that category of respondent. Yet it is not too distant from the truth to suggest that, far from being neutral, such methods of data analysis assume a pluralist society where, within certain normative boundaries, differences of opinion and behavior are acceptable and a reflection of "normal" society (Shaw and Al-Awwad 1994).

Citizens, Service Users, and Social Work Science

Do social work practitioners, service users, policy makers, and citizens have scientific knowledge about social work? Connecting back to the previous chapter we noted Williams and Popay's conclusion that "lay knowledge represents a challenge to the authority of professionals to determine the way in which problems are defined in the policy arena. In this sense it is a *political* challenge to the institutional power of expert knowledge in general, and medical knowledge in particular" (1994, 120). They plead for lay people to have "an active role in the conceptualisation and specification of the nature of the problem, and the design and conduct of the research" and hence for a widening portfolio of acceptable research methods (134). Without wanting to impose negative stereotypes as a way of pleading for some self-evidently superior position, it is reasonably fair to say that there are strong traditions within the professions where the expert does *to* the client what the client cannot do (the expert as operator in a dominance-submission relationship) or where the expert acts as a prescriber who does *for* the client what the client cannot do. While not taking a naïve position (in most instances I am happy for my community physician to be an expert prescriber), there are occasions where the expert as colearner, doing *with* the client what the client can ultimately do for her/himself, is appropriate. This inference also is reinforced by the ways, set out in chapter 7, that social work science has several conjunctions with other forms of citizen knowledge. These make it difficult to weaken, dilute, or even dismiss the knowledge of, for example, service users as nonrational.

Clarification of the issues can be obtained by further developing the part played by experts in science. Evans and Plows raise the question whether it makes sense to speak of the disinterested citizen. They are thinking about deliberative and consultative processes (for example, citizen juries) and want to distinguish between experts, "who may be scientists, activists or others with relevant specialist expertise," and lay citizens or nonexperts, "who have no particular expertise bearing upon the problem beyond that acquired in everyday life" (2007, 828). They thus depart from other positions by arguing for "a more heterogeneous but not unlimited category of expert" (829) while noting in ways not dissimilar from Merton's arguments that experts in one field will be citizens and lay people in relation to other debates.

This poses the question of "how the gap between 'democracy' and 'expertise' can be bridged" (829). Service users, for example, may become expert activists in the field of science as well as social work practice. They then have contact with social work scientists, filter "insider" knowledge back to wider networks, and thus act as "boundary shifters." Hence they are experts by virtue of their experience. This emphasis on continuous engagement also applies to social work scientists who may become acknowledged as experts not because of their qualifications but because of their sustained engagement with a particular topic.

If there are groups of different kinds of experts in both the scientific and the "lay" communities, does this mean that such different categories of expertise are incommensurable? Collins, Evans, and their collaborators seem to think not; nor, at least in some places, does Beresford. "Service users are not suggesting that experiential knowledge is the only knowledge that should be valued or that it should be prioritised. What they have repeatedly expressed concerns about is the way that it has long been systematically excluded from social policy discussions and developments and from social research" (Beresford 2010, 12–13).

But we are still left with a challenge. How do you choose between conflicting expert views? Having more experts does not solve this, but enabling *nonexperts* to be involved, so Evans and Plows maintain, provides scrutiny. They conclude that it is "only those who stand outside the committed knowledge cultures of both the scientific and activist communities who can operationalize a genuinely *civic* epistemology" (2007, 845). Thus they plead that "distinguishing first between experts and non-experts, and then between expert and democratic processes allows debates about controversial science to be analysed in a way that avoids the false oppositions created by the terms 'scientists' and 'publics'" (846).

This argument has implications for how we think about the relationship between social work scientists, practitioners, service users, and citizens. The rather broad strokes in which debates about expertise in social work are conducted tend to ignore the role of the nonexpert. But is there a place for nonexpert involvement in social work science? What sort of social work research issues could this be applied to? Is there a case for citizen juries in social work science, and if so, how could they be selected? How might they be seen as plausibly representative? How would time and training be provided to engage with the social work science community?

Citizen science offers no simple resolution of enduring tensions in social work science, and arguments in favor occasionally risk a somewhat sentimental romanticism. Bloor, in his essay reflecting on citizen science, asks whether the copresence of medical and alternative experts increases clinical effectiveness through, for example, "creative conflict." He refers to his study of South Wales coal miners and miners' lung diseases where it did so prove. "But in the short term this co-presence may simply breed sterile mutual hostility" (2001, 25). He also refers to the influence of AIDS activists on HIV science, alluding to an account by Collins and Pinch (1998a), where subsequently they were absorbed into mainstream culture such that activist representatives were appointed to funding boards and ethics committees and "a process of 'expertification' took place" (Bloor 2001, 27). He concludes pessimistically: "We are back with Alfred Schutz . . . who recognized that the expert would never accept the well-informed citizen as a competent judge of expert performance" (27). Speaking in a medical context, he concludes that "the figure of Citizen Science perches uncomfortably on the consulting room couch," though it still is important to examine, as there "is probably no site in late modern societies where professional and alternative experts meet more frequently than in the consulting room." We need "to problematize rather than sloganize the democratisation of science" (37).

Insiders, Standpoints, and Feminist Science

The territory we have been traversing is sometimes understood in terms of seeing the various participants as either insiders or outsiders and linked with claims to particular kinds of privileged knowledge being associated with being one or the other. It is worth unpacking the kinds of arguments that might be involved in assertions of this kind. First, it is contended that social work research ought to be *for* vulnerable or oppressed participants, and being an insider is the only way one should do research that is for people rather than on or even with people. Second, being an insider enables one to know by virtue of closeness. Outsider researchers in universities or government departments are distant and less able to see. This argument also sometimes has been used to say that qualitative methods are inherently more congenial and consistent with social work values than quantitative methodology. Third, a plea for insider research is sometimes introduced in the guise of an argument about the relationship between theory and practice.

This way of viewing relationships with science is important, even though we have seen enough already to suspect that any cut-and-dried answer is unlikely to persuade all interested parties, as when, for example, participants who are experts in one field are citizens and lay people in relation to another. Similarly, being an insider or an outsider is not an inherent or given status. Hall and—with careful detail—White recorded how they held both insider and outsider roles in relation to their research participants. Hall (2001) "arrived" as an outsider for his ethnography of a young homeless project but became in different ways a partial insider. White (2001) started as an insider for her research within her own social work team yet found herself undergoing a fruitful, if potentially hazardous, process of defamiliarization through which she became in some degree a marginal "inside 'out'" member.

We saw in chapter 7 that Beresford argues that "what distinguishes service user knowledge (or knowledges) and what is unique about it, is that it is based on direct experience" (2010, 12). Are we thus to believe that "insider" knowledge claims by service user and carer researchers, other things being equal, are always (or most of the time) better grounded and more trustworthy than "outsider" knowledge claims? While we have anticipated where we are likely go on this question through earlier discussions of the influence that actors' interests ought to exercise on social work science, the questions of whether particular standpoints have privileged knowledge, and the epistemological status of "insider" and "outsider" knowledge, have made important contributions to social science and social work.

Take, for example, reflections by social work practitioner researchers in a study in Wales.

> Several . . . negative characteristics were . . . mentioned about academic research, including limitations of understanding, experience, and "grasp." Helen viewed practitioner evaluation as more "interactive" and more "valid": "I am doing the job that I am researching. That's the difference—you are actually in the workplace doing the same thing." Sarah also expressed reservations about academic research: "But then I would say perhaps mainly somebody who does it as a job who has not been in the social work field for a while might forget how things work in the real world as well." Jane took this emphasis further and portrayed academic research as more removed from the grass roots: "With practitioner research you live and breathe it. You live and breathe it and you know it so much in depth."
>
> (Shaw and Faulkner 2006, 58)

Consider also example 8.4, where a service user thoughtfully reflects on and theorizes about her experience of social work.

EXAMPLE 8.4
Mrs. Lawton's Theorizing About Social Work

"Mrs. Lawton" is talking about her contact with a social worker based in a London agency, then known as the Family Welfare Association (the direct successor to the London Charity Organization Society), that provided counseling and support for both personal and material problems. She is offering a theory about the way her social worker operated.

"I thought the welfare worker was going to give me some advice as to whether to stay with my husband or leave him. But she didn't give me any advice at all. I think she expected me to keep coming back and by talking about it I would get over it and everything would be back to normal at home. You know what I mean? She gave me the impression that by talking to somebody all my troubles would disappear. Because she kept wanting me to come back, that was the idea . . .

"The social worker asked me what went wrong and I told her. She asked me why does my husband act like he does and what sort of things does he say. And I was giving her answers, but she wasn't giving anything back. Then she would ask me another question. She kept asking me questions, and I would be giving her the answers. I would expect somebody to say, 'Well, why don't you do this?,' or 'Why don't you do that?' . . .

"Once I got talking to the social worker, I felt at ease, but then I realized that she wasn't entering into what I was saying at all. And I thought, 'You are not really listening to me. You are not really interested at all.' She just wasn't giving me an answer or any advice at all. . . . She just kept saying, 'Yes, yes,' in a quiet sort of way and nodding her head and would I like to come back, and that sort of thing."

Source: This material is extracted from Mayer and Timms (1970).

In certain, though not all, respects it is worth recalling how Schutz sees the "stranger." Mrs. Lawton attempts to interpret the cultural pattern of a social group ("welfare workers") she has approached and to orient herself in relation to that group. While commonsense shared understandings typically are good enough to deal with social life, the stranger does not share these assumptions. She becomes someone "who has to place in question nearly everything that seems to be unquestionable to the members of the approached group" (Schutz 1967c, 96). The stranger may be aware that the approached group has a history and culture and that this may be partly accessible to her, but it has never become an integral part of her biography. "Seen from the point of view of the approached group, he is a man without history" (97). The stranger's own home group culture continues to be the scheme of reference for their new environment, and so she or he begins to interpret the new environment in terms of her or his own thinking as usual. In doing so the stranger "is about to transform himself from an unconcerned onlooker into a would-be member of the approached group. . . . Jumping from the stalls to the stage . . . the former onlooker becomes a member of the cast" (97). What was remote becomes proximate. The image of the group from the stranger's home group becomes inadequate, together with the scheme of interpreting it. Yet the stranger "lacks any status as a member of the social group he is about to join and is therefore unable to get a starting-point, to take his bearings." A culture seems to have a unity only for the insider, but "for the outsider . . . this seeming unity falls to pieces" (99). In doing so the stranger is marked by "oscillating between remoteness and intimacy . . . hesitation and uncertainty, and . . . distrust in every matter which seems to be so simple and uncomplicated to those who rely on the efficiency of unquestioned recipes which have just to be followed but not understood" (103–104).

To some readers the foregoing will sound too ready to make concessions. A critical theorist who takes Horkheimer's (2002) stance that in genuine critical thought "both the social structure as a whole and the relation of the theoretician to society are altered, that is both the subject and the role of thought are changed" (211) is likely to retort "I told you so" to Bloor's pessimism. Feminist standpoint theory

> disputes the traditional picture of science by proposing that the mainstream notion of scientific rationality is itself intrinsically masculinist, and thus not amenable to piecemeal correction by feminist scientists. On the contrary, feminists must openly abandon the quest for better "neutral" knowledge, replacing it with a clear emancipatory commitment to knowledge from the standpoint of women's experience and feminist theory.
>
> (McLennan 1995, 392)

Standpoint epistemology starts from Marx's position that a correct vision of society is available only from one of the two major class positions in capitalist society. Hartsock, in a classic paper, reworks Marxist arguments about the division of labor to apply to the sexual division of labor, in which girls define themselves *relationally* while boys do not. She develops her position to claim that these different experiences are replicated in later life as epistemological and ontological differences. This male experience replicates itself in the institutions of class society. Hartsock notes that, given the power of the controlling group to define the terms for the community as a whole, a feminist standpoint is *achieved* rather than obvious. "The standpoint of the oppressed represents an achievement both of science (analysis) and of political struggle on the basis of which this analysis can be conducted" (1983, 288). It is a potentially hopeful position in that "because it provides the basis for revealing the perversion of life and thought, the inhumanity of human relations, a standpoint can be the basis for moving beyond these relations. . . . A standpoint by definition carries a liberatory potential" (289).

Put simply, in response to the patriarchal assumption that women are *less* able to understand, standpoint theory argues that they are *more* able to do so. It does so through two linked assertions: the double vision of the oppressed and the partial vision of the powerful—"privilege and its invisibility to those who hold it" (Swigonski 1993, 174). In a much quoted comment Cook and Fonow say: "The purpose of knowledge is to change or transform patriarchy. . . . Description without an eye for transformation is inherently conservative" (1990. 79).

There are three problems regarding feminist standpoint theory. First, there is an unresolved ambivalence within feminist standpoint theory, especially in regard to criticisms of objectivity (McLennan 1995). Standpoint theory is pulled one way by the realization that, while affirming a radically different agenda for *what counts* as objective, it yet retains a search for universalizing validation, and it is pulled a different way by the emphasis on the diversity of women's experiences. Thus, while Gelsthorpe wishes to retain the terminology of standpoint theory, there is little remaining of its substance when she concludes:

> We cannot assume that black/white, young/old, and so on, experience life in the same way. . . . [We] choose standpoints and standpoints may change over time; they are transitional, not fixed points. . . . This leads me to argue that women do have uniquely valid insights from their vantage points as women, but women are never *just* women. . . . The same goes for men, of course.
>
> (Gelsthorpe 1992, 215)

Second, the argument for a privileged feminist perspective has been criticized from within feminist scholarship. For example, to Hawkesworth the idea "appears to be highly implausible. . . . Given the diversity and fallibility of all human knowers, there is no good reason to believe that women are any less prone to error, deception or distortion than men" (1989, 544).

Third, standpoint theory has been criticized for essentializing the concept of "woman." Third World, black, and lesbian feminisms have together served to "dissolve the conceptualization of "woman" (Olesen 1994, 160), though this in turn creates problems for those who wish to argue that women's experience should have privileged status.

This kind of argument reflects the important influence of the second main position within feminist social work, feminist postmodernism. There is an important feminist postmodern strand in social work, in which the work of Susan Hekman (e.g., 1990) has been especially influential. There has been a move away from standpoint feminism in social work (cf. Orme 2003). While there are important variations within feminist postmodernism, stemming in part from ambiguity about what being postmodern entails, there are several recurring themes.

- A movement away from essentializing concepts of women
- A rejection of objectivism in favor of relativism
- An engagement with pluralist and humanist research methodologies
- A determination to politicize postmodernism and rescue it from any risk of political complacency
- A strong claim that knowledge is contextual and historically specific, and hence a hostility to cross-cultural explanations
- A shift from epistemology and toward discourse and rhetoric

Is it possible to connect ideas of "standpoints" and the questions of whether outsiders or insiders possess privileged knowledge? A helpful perspective on standpoint positions can be achieved by revisiting in detail a classic paper on the sociology of knowledge by Robert Merton. Merton believed that as society becomes more polarized so do contending claims to truth. At its extreme, an active, reciprocal distrust between social groups finds parallel expression in intellectual perspectives that are no longer located in the same universe of discourse. This leads to reciprocal ideological analyses and claims to "group-based truth" (1972, 11). Merton analyzes the relative claims of this nature made by those who are epistemological *insiders* or *outsiders* to the group, in terms we outlined in chapter 2.

Merton develops several criticisms of strong insider positions, but his key point for our purposes is that individuals do not have a single organizing status but a complex status set. "Aggregates of individuals . . . typically confront one another as Insiders and Outsiders" (22). He enters several caveats that enable a reflective assessment of subsequent standpoint positions in social work. He stresses that he is in no way advocating divisions, nor is he predicting that collectivities cannot unite on single issues. Rather, such unity will be difficult and probably not enduring.

Standpoint theory develops this position in an important respect. The "double vision" of the oppressed is in fact an argument for being simultaneously an insider and an outsider. This is the idea that "special perspectives and insights are available to that category of outsiders who have been systematically frustrated by the social system: the disinherited, deprived, disenfranchised, dominated and exploited" (29). The outsider is a stranger. Quoting the early sociologist Georg Simmel, Merton concludes that the objectivity of the stranger "does not simply involve passivity and detachment; it is a particular structure composed of distance and nearness, indifference and involvement" (33).

Social work writers have often failed to distinguish strong and more muted versions of standpoint positions. The latter form of the doctrine claims that insiders (*and* outsiders) have *privileged* rather than *monopolistic* claims to knowledge. This is a position that avoids the erroneous assumption of some radical advocacy researchers who claim that social position wholly determines what understanding is possible. Group identities do significantly influence explanations, but the distinction between tendency and determinism is "basic, not casual or niggling" (27). Merton concludes that, having accepted that distinction, "we no longer ask whether it is the Insider or the Outsider who has monopolistic or privileged access to the truth; instead we begin to consider their distinctive and interactive roles in the process of truth seeking" (36). His conclusion has much to recommend it as a starting point for assessing the relative contributions of insider and outsider models of social work research.

Taking It Further

Reading

Swigonski, M. 1993. "Feminist Standpoint Theory and Questions of Social Work Research." *Affilia* 8(2): 171–183.

Beresford P. 2010. "Re-examining Relationships Between Experience, Knowledge, Ideas, and Research: A Key Role for Recipients of State Welfare and Their Movements." *Social Work and Society* 8(1): 6–21. http://www.socwork.net/sws/article/view/19.

Hawkesworth, M. 1989. "Knowers, Knowing, Known: Feminist Theory and Claims of Truth." *Signs: Journal of Women in Culture and Society* 14(3): 533–555.

Beresford has written extensively, and this downloadable article is representative of his later thinking. His position is generally more radical than that encountered in, for example, U.S. social work. Swigonski gives one of the most articulate social work statements of standpoint theory. Hawkesworth's article is, in my view, unsurpassed.

Cronbach, L., S. Ambron, S. Dornbusch, et al. 1980. *Toward Reform of Program Evaluation.* San Francisco: Jossey-Bass.

I could have included Cronbach following several of the chapters. It perhaps has most to say to chapters 6, 8, and 9. Several of the central themes of this book cannot be adequately assessed without a full reading of Cronbach's luminous, witty, and mind-changing work. I remain perplexed that social work so completely should have neglected his contribution.

Task

Recap the main tenets of each of the four positions in table 8.1.

In discussion with social work colleagues, map examples of social work science onto each of these allegiances.

Tip: a useful single-location resource for this is the four-volume *Social Work Research* (Shaw, Marsh, and Hardy 2015).

[9]
Impacts and Influences

Following reflections on how scientists in social work and other fields have contemplated the challenge of contributing to the application of their work, I consider four questions:

1. How should we think about the uses and misuses of science in social work?
2. What should we make of the demands for the impact of science?
3. Are some forms of social work science less susceptible to influence than others?
4. What is the relationship between knowing and doing in social work science?

I close the chapter, and the book, with a reminder of the limits of social work science.

"No longer can we shroud our citizen-selves behind our scientific subjectivities. We must become scientific citizens" (Greene 1996, 287). In so doing, we might infer, we ought to engage with government, civic society, the media, and those who participate in research and other fields of practice that yield the "science." Collins and Pinch are not obviously preoccupied with applied values of research, yet they complain that "science is often used as a way of avoiding responsibility . . . the substitution of calculation for moral responsibility" (1998, 108). Yet all we have seen through the previous chapters will make us immediately cautious. This is by no means limited to social work or even the social sciences. James Lighthill, the prominent applied mathematician, reflected:

> I don't think I've ever solved the classical problem of the interface between science and politics, though I've given a lot of attention to it. I think it's a peculiarly hard interface. . . . I always felt that there was such an enormous barrier between the way in which politicians look at the world, and the way in which scientists do, that it's very hard to penetrate it. And yet it must be penetrated.

The main problem, he suggests, is the difference in mode of thought. Scientists "are obsessed with truth and accuracy, and also concerned with the long processes in time by which truth is discovered. Politicians want answers very quickly, and they're more concerned with how things are presented to the electorate" (Wolpert and Richards 1997, 66).

Downstream consequences also are unpredictable. Cronbach and colleagues tell an anecdote about how a study that showed the lack of success of spelling drills led to the perverse irony of more spelling drills, concluding: "Here we glimpse a significant generalisation: whether an evaluation is launched to promote a cause or to report neutrally on events, the measurement procedures and reports can have a wholly unanticipated influence on what happens next" (1980, 27).

Hearing the Siren Voices for Science Usefulness

In the face of such experience we may concur with Eleanor Chelimsky, formerly the assistant comptroller general for program evaluation and methodology in the U.S. General Accounting Office, who cautions we should not become overly preoccupied with models of research use. Chelimsky believes "it is often the case that . . . evaluations are undertaken *without any hope of use.*" Expected nonuse is characteristic of some of the best evaluations, including "those that question widespread popular beliefs in a time of ideology, or threaten powerful, entrenched interests at any time" (1997, 105). Thus, "there are some very good reasons why evaluations may be expert, and also unused" (105). Her comments are both sane and plausible: "To justify all evaluations by any single kind of use is a constraining rather than an enabling idea because it pushes evaluators towards excessive preoccupation with the acceptability of their findings to users, and risks turning evaluations into banal reiterations of the status quo" (106).

Yet the expectation that social work science should be of social, policy, and political value remains unflagging. It feeds advocacy for:

1. Changed forms of scientific work
2. Active commitment to understanding and promoting knowledge utilization
3. Calls for engagement

By way of illustration of the first of these, the prefix "trans-" figures prominently in discussions of scientific activity in social work. It may be "transdisciplinary," as a way of referring to spanning unhelpful boundaries (one of the worst accusations to receive is about working in a silo); or "translational," when calling for engagement and bridging the knowledge and practice gap; or "transformative," focusing on impact, change, and end-user benefits (cf. Nurius and Kemp 2012).

Such pleas sometimes sound too forgetful of what Stephen Fuller calls "the increased disciplinization of the scaled-up modern university," where the autonomy of inquiry is "relativized to particular disciplines" (2009, 23). With disciplinization comes specialization. This is not new. As early as 1919 Weber, when speaking to graduate students in Germany, referred to the "much talked-about" issue of how "science has entered a phase of specialization previously unknown" (1948, 134) and how the individual achieves by being "a strict specialist." Medawar (1984) responds partly to counter the argument that science has become ever more unintelligible to the lay person through specialization. He thinks this is overstated. The *ideas* of science are often quite simple, but the means are harder—"it is scientific performance rather than the scientific conception that tends to bewilder the lay public" (7). Fuller's prescription is to present the task of education as to release the specialist insights of research into "larger social settings, and not to reinforce their original theoretical packaging by treating students as if they were potential recruits to specialist ranks" (30).

On the second response, gradually increasing pressure from Western governments for social research to have use-value, along with empirical studies of how scientists work, have turned eyes away from relatively passive ideas of "dissemination" toward more active, participative notions of science "utilization" (Walter et al. 2004; cf. Ruckdeschel and Chambon 2010). The distinction sometimes is implicit rather than spelled out. For example, in a valuable exploration of how we may theorize science dissemination Dearing suggests that "dissemination embeds the objectives of both *external validity*, the replication of positive effects across dissimilar settings and conditions, and *scale-up*, the replication of positive effects across similar settings and conditions" (2009, 504). He distinguishes diffusion, by which he means broad-based adoption, from dissemination—the extent and quality of response to the diffusion. "Diffusion occurs through a combination of (a) the need for individuals to reduce personal uncertainty when presented with new information, and (b) the need for individuals to respond to their perceptions of what specific credible others are thinking and doing, and (c) the general felt social pressure to do as others have done" (506).

He refers to the work of Katz, who proposed that diffusion occurs more readily when the innovation matches the characteristics of "the pensive adopter" in terms of its "*communicability* (the degree the which an innovation's utility is easily explained), *pervasiveness* (the degree to which the innovation's ramifications are readily apparent), *risk* (the degree to which an innovation is dissimilar to what it replaces), and *profitability* (the degree to which an innovation is perceived as more cost effective than alternatives)" (Dearing 2009, 510). Katz thought of these dimensions collectively as constituting an innovation's compatibility with its context, stressing the kinds of accommodation that occur between innovation and context.

The third response, that of calls for mutual engagement between academics and those who seek use-value from science, especially governments, has been heard from all quarters, including, for example, the National Institutes for Health and, in the United Kingdom, the Council for Science and Technology. The relationship typically is seen as difficult. Since the later nineteenth century the growth of the social sciences and the expansion of the state have been very closely linked. But among social scientists "this long and close association with public policy making continues to generate tensions, misgivings and self-criticisms about the ethics of knowledge development with social and political consequence" (Bastow et al. 2014, 141). Liberal democracies embed in the nation-state the idea of a public interest that has a close fit with an academic culture that "sees impartial service to knowledge and fostering the 'public interest' as the twin prime rationales for researchers' relative autonomy and independence from public pressures" (143). But this is offset by the fact that this takes place in a heavily politicized environment. Hence "to serve the public interest but not to let social science become political" (143) embraces the central tension. Layered over this, leftist or liberal administrations have different sets of priorities and indeed hangups compared with neoliberal or right-wing administrations.

The Council for Science and Technology highlighted four inhibitors as to why engagement between academics and policy makers was not as strong as it could be, in terms that no doubt apply beyond the United Kingdom (Council for Science and Technology 2008).[1]

EXAMPLE 9.1
Inhibitors to Engagement

1. Less-than-professional working relationships. "Both sides need to better understand the constraints that the other is working under and exactly what each requires. When Government consults academics it needs to be a more 'intelligent customer.' . . . In response academics need to ensure they fully understand what officials are looking for" (8).

2. Ignorance on both sides of what good engagement can deliver. There needs to be individuals and structures in the policy officer community who have sufficient understanding of the academic world. Universities and academics need to be proactive at promoting their work to government.

3. Mistrust between academics and government. On the part of academics the prime cause is a sense in some cases that they are being pressed to provide "policy-based evidence" and to work under short time scales. But also "academics must recognise that where a particular view does not prevail, or where decisions are taken for political reasons, this does not mean the academic input was not valued" (10).

4. Failure to value the relationship. Engagement sometimes was felt by each side to be a token process to "tick the engagement box."

Source: Council for Science and Technology (2008).

The authors of the report concluded uncompromisingly that "the central message is that both academics and policy makers need to alter their behaviour to overcome the barriers" (8).

Policy language is, of course, rhetorical, although this does not imply that it is thereby "false." In addition, government policy positions are rarely "fixed." Priorities at the operational level move in and out of focus, and midlevel government ministers change. The policy community is not homogenous and may be marked by major changes, whereby ministers quickly move on and officials become marginalized or move on to greater things. These processes may not be well understood by the academic community, and when they *are* there is the risk that they become co-opted into the policy community.

Cronbach relates this to how the governance system uses information. Governance has "a context of command and a context of accommodation" (Cronbach et al. 1980, 83–84). However, "most action is determined by a pluralistic community not by a lone decision maker" (84). Decisions are not "made," but they emerge. There is a "drift toward decision" through processes of accommodation (cf. Weiss 1980). Cronbach is not happy with the way the summative (how good is a service?)/formative (how can a service be improved?) distinction is used. "As we see it, evaluations are used almost entirely in a formative manner when they *are* used" (62). They go for *future* use as the value. "Far more is to be learned from evaluation than a precise answer to an obsolete question" (64). They offer a framework for thinking about how evaluation use works by taking into account the relationship between the level of community agreement

	Consensus with regard to values	Disagreement
Consensus with regard to facts	Rational analysis	Compromise
Disagreement	Judgment	"Inspiration"

Figure 9.1 Agreement on Facts and Values
Source: Cronbach et al. (1980).

on values and the level of agreement about the facts of the matter (figure 9.1).

Van de Ven and Johnson (2006) offer one of the more articulate pleas for an approach in which applied researchers collaborate with research subjects (that is, clients) to identify the questions, methods, and outcomes. In their plea for engaged scholarship they set out a research strategy that negotiates the gap between theory and practice in a way that ought to be applicable to social work science. Their approach navigates the problems that arise when "theory" and "practice" are seen as separate entities and when science application is seen in simplistic knowledge transfer forms.

Uses and Misuses and Social Work Science

Lines of reasoning of these kinds take us only so far. It is inadequate to speak of social work science and its use (or misuse) as if it was something homogenous. We think of social science in terms of whether it is applied or not—whether it is undertaken with the intention of being fairly directly helpful in some way for some "social problem." The ideas of "social problem" and "applied science" both call for elaboration.

Social Problems

While Merton's modified functionalist understanding of social problems is inadequate as an overall analysis (cf. Timms 2014), his conceptual elucidation is of value (Merton 1971). Asking what social problems are, he says there at least six connected questions to answer:

> (1) The central criterion of a social problem . . . (2) the sense in which social problems have social origins; (3) the judges of social problems, those people who in fact principally define the great problems in a society; (4) manifest and latent social problems; (5) the social perception of social problems; and finally (6) the ways in which the belief in the corrigibility of unwanted social situations enters into the definition of social problems.
>
> (799)

Given the absence of consensus, one group's problems will be another group's assets. For one group something only begins to be a social problem if action is taken to enforce the position of the other group. Take, for example, the Obama healthcare reforms and many other comparable instances. He insists that the existence of disparate values and interests does not "dissolve the concept of social problems in the acid of extreme relativism" (805).

He avoids rationalism of the kind that assumes once problems are revealed they will be abandoned. "The sociological truth does not instantly make men free" (807). But the sociologist, as sociologist and not as citizen, "takes a distinctive and limited part" in social controversy. "In his capacity as sociologist, emphatically not in his capacity as citizen, the student of social problems neither exhorts nor denounces, neither advocates nor rejects. It is enough that he uncovers to others the great price they sometimes pay for their settled but insufficiently examined convictions and their established but inflexible practices" (808–809). His view of the role of the sociologist includes rejecting that "insolent arrogance that would have us pretend to know that society is bound to move in the one direction of cumulative improvement or in the other of continuing decline" (809).

For something to be seen as a social problem in Mertonian functional terms it must be perceived as corrigible. Put differently, the value orientations that people hold toward the preventability or controllability of unwanted social conditions will affect the perception of social problems. At the extremes, there are those societies where problems are viewed in a fatalist light. A telling example of this is given by Riessman (2001) in her narrative analysis of a woman attending an infertility clinic in Kerala, in southern India. "In such a society, the social problems are chiefly or altogether latent" (Merton 1971, 814). At the other extreme are those societies committed to an activist philosophy of life "that takes just about everything in society as being in principle subject to human control." Such a society "may have many manifest social problems though fewer problems altogether" (814), though societies rarely are at one or other extreme.

Applied Science

If the notion of a social problem has diverse strands, the assumption that good social work science will always be of direct or indirect value also calls for inspection. To cross a bridge from Merton, he recognizes an intellectual division of labor in science generally, rather than an all-or-nothing commitment to either pure or applied work. Some are suited to the exclusive pursuit of one or the other, "some may move back and forth between both; and a few may manage to tread a path bordered on one side by the theoretical and on the other by the practical or applied" (Merton 1971, 793).

We may think of social work as part of a wider "family" of fields, including health, management, education, criminology, and perhaps social policy, that are "applied," in contrast to social science fields such as sociology, much of psychology, and some of economics and politics, which are more in the nature of a "discipline." A different but logically rather similar way of thinking is to view each of these fields as having "pure" and "applied" "ends." This reflects how we think of fields like psychology and economics (as well, of course, as very different fields, like mathematics).

We tend to think of basic and applied research either as two separate categories or as being on a continuum, such that one cannot move closer to one end without moving further from the other. In contrast to either of these ways of placing social work science I suggest, first, that the pure (or basic) and applied categories, while different, are not as far apart as we are accustomed to think, and second, that the way we think about "applied" has been too influenced by a rather instrumental notion of use, quality, and value.

In the United Kingdom in the 1990s there was an introduction of bridging categories like "strategic research"—that is, science that has use in mind but not as an identifiable application or product. This way of thinking also influenced work in the United States during that period. Ideas such as "purposive basic research" came into the language, denoting fundamental research done with a general application in mind or basic research that is "mission oriented." A common response to such thinking is to take the single dimension pure/applied and place strategic science somewhere in the middle, although this introduces the problem of fuzziness at the boundaries.

A different kind of criticism is that it attempts "to force into a one-dimensional framework a conceptual problem that is inherently of higher dimension" (Stokes 1997, 71). *Pasteur's Quadrant* was the title of an influential book by Stokes. His premise there is that it is possible to gain a very different view of the relationship between basic and applied research when we consider how research decisions made in some very important studies have been guided by both basic *and* applied considerations. Stokes takes Pasteur's work on microbiology. "As Pasteur's scientific studies became progressively more fundamental, the problems he chose and the lines of inquiry he pursued became progressively more applied" (13). Stokes includes the social sciences. For example, the economist Keynes "wanted to understand the dynamics of economies at a fundamental level. But he also wanted to abolish the grinding misery of economic depression" (17). Pasteur belongs not halfway along but *at both ends*. His work could not be placed on a single dimension whereby more of one means less of the other: it was fully applied and fully basic.[2] For Stokes there does not exist a category of science to which one can give the name applied science. There are sciences and the applications of science, bound together.

This immediately makes us think about the question in a different way. Instead of asking whether science is aimed at being useful *or* at gaining greater understanding and knowledge, we now ask *two* questions. Does the science have considerations of use? Does it seek fundamental understanding? We can cross the answers to yield four possible positions, as in figure 9.2. Helpfully complicating ideas of applied social work science provides a grounding for appreciating different kinds of use.

Fundamental understanding?	No	Yes
Yes	Pure basic research (Bohr)	Use-inspired basic research (Pasteur)
No	Pp	Pure applied research (Edison)

Figure 9.2 Pasteur's Quadrant
Source: Stokes (1997).

Uses of Social Work Science

Carol Weiss delineated the political contexts in which scientific work is located. Although she was primarily concerned with evaluation and policy research at the federal level, her empirical work with policy and program staff resonates more widely with social research. She also exposed the limitations of conventional instrumental views of the political use of information, through her conceptualization of use as enlightenment. Furthermore, Weiss imbued models of use with a realistic view of the public interest. "More than anything she has struggled towards a realistic theory of use. These shifts started a debate in evaluation that goes on to the present day about the role in evaluation of idealism and pragmatism" (Shadish et al. 1990, 207–208).

Recalling the warning that to justify all science by a single kind of use is a constraining rather than an enabling idea, the conventional rational assumption about the utilization of science was that research led to knowledge, which in turn provided a basis for action of an instrumental, social engineering kind. The historical roots of this view are deep. Weiss was not the first to question the legitimacy of this view, but her significance lay in the empirical underpinnings, explanatory cogency, and plausibility that she conveyed. With her colleagues she interviewed 155 senior officials in federal, state, and local mental health agencies. Officials and staff used research to provide information about service needs, evidence about what works, and to keep up with the field. However, it was also used as a ritualistic overlay, to legitimize positions, and to provide personal assurance that a position held was the correct one. At a broader conceptual level it helped officials make sense of the world. For all these purposes, "it was one source among many, and not usually powerful enough to drive the decision process" (Weiss 1980, 390). As for direct utilization, "instrumental use seems in fact to be rare, particularly when the issues are complex, the consequences are uncertain, and a multitude of actors are engaged in the decision-making process, i.e., in the making of policy" (397).

Research use was also reflected in officials' views of the decision-making process. Decisions were perceived to be fragmented both vertically and horizontally within organizations and to be the result of a series of gradual and amorphous steps. Therefore, "a salient reason why they do not report the use of research for specific decisions is that many of them *do not believe that they make decisions*" (398). Hence the title of her paper—"Knowledge Creep and Decision Accretion." This provided the basis for her argument that enlightenment rather than instrumental action represents the characteristic route for research use.

Weiss's position was challenged especially by Patton. His argument was twofold (e.g., Patton 1988, 1996). First, he reasons that Weiss wrongly generalized from policy research to other kinds of science knowledge. "It makes sense that policy research would be used in more diffuse and less direct ways than program evaluation" because they are "different kinds and levels of practice" (1988, 12). Second, her vision is "quite dismal." "The Weiss vision, in my judgement, is not marketable" (11). Her response is to complain that in Patton's world "everybody behaves rationally" (Weiss 1988, 18). She also elaborates her earlier argument regarding the politics implicit in research. We are familiar with the arguments for ways in which the political process may act in negative ways. They include:

- Pressures to limit the scope of inquiry
- Demand that academics meet unrealistic timeframes
- Indirect pressure to distort the study results through requests that alternative interpretations of the data are considered
- The selective dissemination of results
- The suppression or critical delay of publication of the report

Yet there are ways in which politics can provide paradoxical support for useful science. Commissioning of scientific inquiry often functions as a solution to political disagreement. While this may itself be a negative consequence, it can have positive effects. While it would be reassuring to believe otherwise, it is likely that much scientific inquiry is initiated to confirm existing beliefs or a policy position. But such delaying tactics may unwittingly feed longer-term enlightenment uses of science.

The straightforward distinctions between instrumental and enlightenment uses of science is elaborated by Kirk and Reid (2002). They suggest a six-fold distinction between:

- Instrumental
- Enlightenment
- Conceptual use—to influence thinking
- Persuasive use
- Methodological utilization
- Indirect use—use that is mediated through, for example, research-based models

They also helpfully underscore that we do not know enough about the actual utilization patterns that take place. "The bottom line for research utilization is what actually happens in the field among practitioners" (194). In the context of such elaboration, an ambitious effort at the framing of research use comes from the Social Care Institute for Excellence (SCIE), in the United Kingdom. A SCIE report on improving the use of research in social care practice (Walter et al. 2004) addresses the status of research use in social care and social work and focuses on furthering efforts in this regard. While not prescribing or endorsing any particular research methodology, the report concludes that there are three distinct models of research use in social care.

1. Research-based practitioner
2. Embedded research
3. Organizational excellence

The first model is a linear process drawing on practitioner responsibility and autonomy and requiring professional education and training. The embedded research model is achieved by rooting research in the systems and processes of practice, thus resting responsibility with policy makers and managers, also as a linear and instrumental process. The organizational excellence model calls for research leadership and the development of a research-minded culture not unlike that advocated in the learning organization literature. Ongoing learning and the adaptation of research utilization within organization are also required, perhaps through partnerships with universities and other organizations (Walter et al. 2004; cf. Ruckdeschel and Chambon 2010).

The implicit distinction between models that start from the practitioner and those that begin with the organization is reflected in the work of the sociologists Bloor and McKeganey. Bloor argues that "the real opportunities for sociological influence lie closer to the coalface than they do to head office, that the real opportunities for sociological influence lie in relations with practitioners, not with the managers of practice" (Bloor 1997, 234).

In reviews of a street ethnography of HIV-related risk behavior among Glasgow male prostitutes and comparative ethnographies of eight therapeutic communities, he suggests there are two ways in which ethnography might speak to the practitioner. First, it may "model" a service delivery that can be transferred to service providers. For example, "ethnographic fieldwork, in its protracted and regular contacts with research subjects, has much in common with services outreach work" (227). This point has some similarities with the argument of reflective evaluators that ethnography may be colonized and translated in ways that may be utilized in direct practice. From the therapeutic community studies, Bloor suggests that the very act of comparative judgment can model helpful service practice. "Rich description of particular kinds of therapeutic practice can assist practitioners in making evaluative judgements about their own practices" (229). Second, ethnographers may, where appropriate, draw practitioners' attention to practices they think worth dissemination and further consideration. Bloor and McKeganey (1989) list seven practices that seemed to them to promote therapy in their original settings and that they discussed with the practitioners in the therapeutic communities. Incidentally, they point out the corresponding implications for ethnographers, in that "any attempts to further exploit the evaluative potential of ethnography for a practitioner audience must be paralleled by a growth in ethnographic studies which focus on practitioners' work, not practitioners' conduct" (210).

Until that time when "citizens themselves commend the work of practitioners . . . it is not the place of sociologists to murmur of false consciousness and demand resistance to pastoral care" (Bloor 1997, 235). In a deceptively gentle later piece he concurs with and illustrates that the researcher has an obligation to bring about good. "It is an obligation we all share in all social settings and that therefore stretches across the entire duration of a research project. And it is not an obligation we can ignore with impunity in the service of some higher calling such as scientific rigour" (Bloor 2010, 20).

Misuses of Science

There some misuses that in the eyes of some barely count as such. Carl Djerassi, for example, in his novel *Cantor's Dilemma* (1989), is writing about "elite" scientists' behavior and culture and "that the drive to succeed means that the elite is necessarily blemished" (Wolpert and Richards 1997, 14). He does not think there are many scientists who if honest would say they do science for science's sake—"if you did you'd be happy to publish anonymously" (14).

A different kind of probably frequent misuse is countered by Merton when he says: "There is no merit in escaping the error of taking heterodoxy to be inevitably false or ugly or sinister only to be caught up in the opposite error of thinking heterodoxy to be inevitably true or beautiful or altogether excellent" (1971, 832). Romm's advice may seem unduly optimistic when she advises: "It is incumbent on the knower to be aware continually of partiality, and it is incumbent on the knower as intervener to attempt to instill such an awareness in the consciousness of participants (by requiring them to listen to the position of 'the enemy')" (1995, 163).

More obviously, history is scattered with instances showing that at their worst "scientific concepts can reinforce a vast array of dangerous or hateful political and moral agendas" (Jacob 1992, 495). "There are evil ends directing actions, and there are ignoble curiosities of the understanding" (Merton 1971, 794). Max Born, the German Nobel Prize–winning physicist, writing in 1946 in the immediate aftermath of the war, reflected, "We have now a terrible responsibility. We should do nothing without thinking where it may lead to, and we cannot retire to an 'isolationism' or an ivory tower. Yet I am quite convinced that the eternal value of science lies in things remote from any applications, good or bad, in finding the truth about reality" (Greenspan 2005, 267).

We perhaps smile at Charles Babbage's (1830) suggestion that "there are several species of impositions that have been practised in science, which are but little known, except to the initiated. . . . They may be classified under the headings of hoaxing, forging, trimming and cooking." He defined hoaxing as "deceit [which] is intended to last for a time, and then be discovered, to the ridicule of those who have credited it" and trimming as "clipping off little bits here and there from those observations which differ most in excess from the mean, and in sticking them on to those which are too small." He added cooking, defined as "an art of various forms the object of which is to give ordinary observations the appearance and character of those of the highest degree of accuracy. One of its numerous processes is to make multitudes of observations, and out of those select those only which agree, or very nearly agree. If a hundred observations are made, the cook must be very unlucky if he cannot pick out fifteen or twenty which will do for serving up" (7). Finally, forging was done by "one who, wishing to acquire a reputation for science, records observations which he has never made." We likewise may smile at G. K. Chesterton's remark that "it is perfectly obvious that in any decent occupation (such as bricklaying or writing books) there are only two ways (in any special sense) of succeeding. One is by doing very good work, the other is by cheating" (quoted in Bastow et al. 2014, 37).

Medawar suggests a view of fraud that derives from the community characteristics of science. The principal cause of fraud is "a passionate conviction of the truth of some unpopular or unaccepted doctrine . . . which one's colleagues must somehow be shocked into believing" (1984, 32). He gives the example of the notorious Sir Cyril Burt's work on the IQ of twins, to stress heredity. Why was he not exposed earlier? Because "Burt told the IQ boys exactly what they wanted to hear," and colleagues suffered "sheer stupidity. They are almost unteachable too" (33). Jacob is nonetheless probably right when she concludes that

> Science can be socially framed, possess political meaning, and also occasionally be sufficiently true, or less false, in such a way that we cherish its findings. The challenge comes in trying to understand how knowledge worth preserving occurs in time, possesses deep social relations, and can also be progressive . . . and seen to be worthy of preservation.
>
> (1992, 501)

"Impact"

There have been loud claims as to the importance of planning for and prioritizing research impacts. The National Institutes for Health site is replete with references to "impact." It has come to the fore in the United Kingdom as much as elsewhere. The UK funding councils define impact as "the demonstrable contribution that excellent research makes to society and the economy."[3] Key aspects of this definition of research impact are that impact must be demonstrable and that one cannot have impact without excellence. The research councils express it as follows:

> We aim to achieve research impact across all our activities. This can involve academic impact, economic and societal impact or both:
>
> - *Academic impact* is the demonstrable contribution that excellent social and economic research makes to scientific advances, across and within disciplines, including significant advances in understanding, method, theory and application.
> - *Economic and societal impact* is the demonstrable contribution that excellent social and economic research makes to society and the economy, of benefit to individuals, organisations and nations.

They distinguish the impact of social science research as:

> 1. *Instrumental*: influencing the development of policy, practice or service provision, shaping legislation, altering behaviour
> 2. *Conceptual*: contributing to the understanding of policy issues, reframing debates
> 3. *Capacity building*: through technical and personal skill development

Determining the impact of social science research is not a straightforward task. Policy and service development is not a linear process, and decisions are rarely taken on the basis of research evidence alone. This makes it difficult to pin down the role that an individual piece of research has played. In a blog post two historians criticize what they see as the "essentially paternalistic, top down approach to assessing whether a project will have value and relevance."[4]

> In resisting being told how to do things from those above us within the university, research councils and government, we should also be wary of doing the same when we're working with those outside the academy. . . . It is not the exchange of knowledge that is most effective in this case of a history or arts and humanities project, but the use of conversations to question the very foundations of how we practise within a professional sense.
>
> (King and Rivett 2013)

They detect a general paradox that "the current focus on achieving research impact that offers new possibilities for collaboration, and has allowed us to pursue our individual projects, can also limit innovation. . . . It leads us into a focus on a more one-way form of dissemination from research to public(s) and those parts of partnerships that are measurable in terms of impact."

The most significant study, hitherto, of the impact of science was completed by a team at the London School of Economics (Bastow et al. 2014). Based on a three-year research project studying 370 UK-based academics, their concern is with the dynamics of external influence.

> Two broad narratives regularly merged. The first perspective was that because it takes a lot of time and effort to "translate" academic work for audiences outside higher education, and even more to get it noticed or accepted by significant decision-makers, this was increasingly a specialized academic role. . . . The second perspective was that in any discipline or institution the same sets of people tend to be more efficient and effective than their colleagues, across all aspects.
>
> (Bastow et al. 2014, 35)

Both narratives had caveats. For example, on the specialist role argument, it was acknowledged that "a scattering of other researchers will regularly 'get lucky' when their work happens to strike a chord" (35). On the second perspective, those who took this view recognized that not all good academics are good communicators, although "there should be a strong overall correlation between the quality and frequency of researchers' publications, academic ranks and reputation and their external visibility and persuasiveness" (36).

They conclude: "Both of the two popular (or conventional) views of how social scientists acquire external visibility are partially right and partially wrong" (63). There is a substantial group of academics who are "fairly single-mindedly pursuing academic influence alone, with low external visibility" (63), but there was no obvious group who specialized in being close to the customer. But the view that there are elite academics who predominate in external visibility also gained little support. Medawar differentiates scientists still further. Scientists include "collectors, classifiers and compulsive tidiers-up; many are detectives by temperament and many are explorers; some are artists and others artisans. There are poet-scientists and philosopher-scientists and even a few mystics" (1984, 10–11).

Stepping back from the extensive empirical details that their study yielded, Bastow and colleagues introduce the notion of a Dynamic Knowledge Inventory (DKI), which "sums complex processes by which professions, institutions and knowledge communities accumulate, store, re-package, and re-deliver knowledge, often over long time periods" (Bastow et al. 2014, 254). Knowledge falls in and out of use in the DKI and consists of all kinds of knowledge—basic, tacit, commonsense, applied, experiential, expert, and so on. In terms that resonate with the way science knowledge has been understood throughout this book, they conclude: "It seems likely that many forms of knowledge are important for societal guidance, and that the neglect of any form can potentially have deleterious consequences" (258).

Collins and Evans (2008) have proposed an illustrative fifty-year rule: scientific disputes take a long time to reach consensus, and thus there is not much scientific consensus around; and the speed of political decision making usually is faster than the speed of scientific consensus formation, and thus science can play only a limited part in decision making in the public domain. DKI is a collective process. "It is not something easily owned or controlled by any actor or coherent set of actors. Indeed, there is often competition" (Bastow et al. 2014, 254). Also, "at any point in time, much of the value of the DKI must be latent" (258).

Journal Impact Factors

It is not possible to discuss the impact of science and research without saying something about the growing and far from simple field of journal impact measures. There is an increasing number of systems for measuring journal impact, although the frontrunner at the time of writing is still

TABLE 9.1
Extracts from Thomson Reuters Journal Citation Report for "Social Work," 2013

RANK POSITION	JOURNAL TITLE	TWO-YEAR IMPACT FACTOR
1	*Trauma, Violence, and Abuse*	2.939
2	*Child Maltreatment*	2.706
3	*Child Abuse and Neglect*	2.135
20	*Social Service Review*	0.791
27	*Social Work Research*	0.535

Thomson Reuters Journal Citation Reports. Table 9.1 gives selected entries among the thirty-nine "social work" journals for 2013, listing the rank order, the journal title, and the two-year impact factor.[5] My purpose in selecting these titles is to make general points about what measurements of this kind gauge and what they seem to mean, rather than to comment on any particular journal.

Drawing on table 9.1 and from my experience as a journal editor who has been through the prolonged process of securing admission to this index, the following observations seem reasonable. First, were we to compare scores for social work journals with those for, for example, medicine or the traditional sciences, the former's scores are relatively very small. Second, while we have no direct evidence of the respective weight of honorific and critical citations, any citation, positive or negative, counts toward the score. Third, citations tell us nothing about the impact of articles on readers rather than writers. It is possible that *download* figures tell us more about reader interest than *citations*. Fourth, rankings reward established journals and exclude new ones. Linked to this, fifth, what kinds of articles most get cited? My suspicion is that genuinely innovative work that may fit nowhere in how social work is currently understood and taught—and sadly there is all too little of it in social work (Phillips and Shaw 2011)—is less likely to get cited. Impact targets push both editors and publishers to try to secure citable articles, and it is probable that the most heavily cited articles are review pieces rather than original empirical work. This is the familiar point we know about measurement-led indicators: they alter behavior. Finally, in the social sciences, including social work, impact scores may on average be falling.

On the journals listed in table 9.1 it is apparent that the top-ranked journals fall in the same broad field and probably have associations with medicine and health science interests. Also, although the two U.S. social work journals both fall below the midpoint in the rankings, this may not convince readers that this reflects a real measure of importance.

Yet despite serious reservations regarding such indices of the worth of science in social work (cf. Blyth et al. 2010), they have a dangerously seductive quality. A recent editor of a major social work journal, speaking in an as yet unpublished research project, remarked:

> At the moment officially the impact of the journal doesn't necessary matter, but I find that a bit hard to believe really . . . [so], much as we are skeptical about impact factor and league tables . . . at the end of the day they are the rules of the game and if people want to prosper in the game they have to play by the rules, whether they think they're crazy or not.

To all this I find myself wanting to say, "No—and yet . . ." I feel uncomfortable with the language of "impact." I share the position of someone speaking as part of a study to understand the kinds and quality of social work research, when s/he said, "If it's methodologically poor research that has a large impact then I would judge it as not useful because it's actually influenced moves in the wrong direction. It's added to confusion and misunderstanding and bad policy rather than the reverse" (Shaw and Norton 2007, 45).

"Impact" is, though we rarely acknowledge it, a metaphor, taken primarily from the field of dynamics, to refer to the striking of one body in momentum against another. Being "hit" by research is perhaps too common an experience. I am also disquieted by the rush to measure—to reduce, to simplify. This leads, as the sociologist Ben Baumberg observes,[6] to "untruthful truth," such that no one believes the impact scores but everyone uses them. This sense of seductive madness is worth consideration. It promotes what Baumberg calls *unhelpful reflection*—reflection that is not helpful for application.

Are Some Forms of Social Work Science Less Susceptible to Influence Than Others?

We have seen that probably there is a substantial group of academics who are "fairly single-mindedly pursuing academic influence alone, with low external visibility" (Bastow et al. 2014, 63). But are there some forms of social work science that intrinsically are less likely to yield payoffs at the various levels on which social work operates? The question is extensive, albeit one to which we have implied answers in earlier chapters. At this stage I restrict myself to making a distinction between the broad kind of social theory that is brought—implicitly or explicitly—to scientific work and the implications of different general methodological stances.

The *social theory* distinction I have in mind was made in 1936 by Stuart Chapin, one of the various significant figures on the borders and sociology and social work, in his presidential address to the American Sociological Society, which was published as the opening article in the first issue of the newly established *American Sociological Review*. He distinguishes "planned social action directed towards goals" and "social action that flows from unintended consequences of inter-relationships among personal social forces" (Chapin 1936, 1), or, as he more succinctly expresses it, "the unplanned combinations of independent individual behaviors"—"impersonal social forces" (2).

He argues that there is a social theory counterpart to this dichotomy—what he calls normative social theory and non-normative social theory. Non-normative theories do not try to say what should be done but what and how things occur and sometimes why. By contrast,

> in normative social theories we encounter value judgments that stress differences in kind. In the non-normative social theories we more frequently encounter quantity judgments that stress differences in degree within each kind. . . . The mental set of the normative theorist is to explain results by treating impersonal consequences as if they were ends or goals. On the other hand . . . the mental set of the non-normative social theorist is to abstract impersonal principles from social situations that were essentially personal in origin. . . . As a result of these differences, normative social theory is able to explain planned social action better than it can explain the unplanned results of the interrelated but independently planned social actions.
>
> (4)

A more frequently encountered form of the argument for different potentialities of social work science for application and external influence is that qualitative *methods* are less amenable to policy and intervention applications than more structured and quantitative methods. This may be seen as simply a characteristic of such methods or sometimes as the failing of qualitative researchers to tackle significant questions. There are plausibly relevant factors that bear on this proposition. First, from the perspectives of governments the utility of statistics and politically neutral "facts" renders statistical and survey-based reports attractive. Second, research often has entailed a social engineering model aimed at social justice through belief in rigor and the "inevitability of gradualness." This tradition inherited the assumption of politically unproblematic facts which would speak for themselves. Third, discipline developments over the last one hundred years led in the United States to social work developing in separate university locations from, for example, sociology. When qualitative methodology permeated sociology, it was usually without a policy focus. This facilitated an antiquantitative thrust in sociology. While "applied" traditions developed rather differently in U.S. and UK social science (cf. Jacob 1987; Atkinson, Delamont, and Hammersley 1988), large sections of sociology came to have a well-nigh universal distaste for social reform. Taken together, these points help explain the perceived grounds for the criticism, sometimes heard, that qualitative research yields little of applied value.

However, Hammersley (2000) notes that there is a model of qualitative research that provides a way to think about the relevance of qualitative research.

EXAMPLE 9.2

Hammersley on the Relevance of Qualitative Research

Hammersley notes a model of the practice/research relationship behind criticism of the relevance of qualitative research that sits uneasily with the assumptions about the nature of the social world of qualitative research. The assumption is of an engineering model of providing techniques that work. The engineering model "implies that research findings have inherent and determinate practical implications" (393)—or at least should have. What is likely to be a more appropriate understanding is an enlightenment model that treats findings as more uncertain and unpredictable research and as "providing resources that practitioners can use to make sense of the situations they face and their own behaviour, rather than telling them what is best to do" (93).

In developing this case he draws on "the fundamental insights of interactionism and phenomenology; notably, that the social world is complex and processual in character, so that there is a high level of contingency inherent in any course of action" (400). He borrows from an earlier article to identify five different capacities that qualitative research brings for contributing to practice.

1. Appreciative: "The ability to understand and represent points of view which are often obscured or neglected" (394–395). A problem with mainstream approaches to research use is that they are too closely aligned to a "correctional" perspective, where the perspectives of the policy enforcement community come to dominate. This term suggests a helpful connection between research a generation or more ago on deviance and how we should address research use in social work.

Qualitative researchers are sometimes accused of taking sides. But "while partisanship is undoubtedly a danger in appreciative research, it is not automatically built into it" (395). "By contrast, partisanship is built into correctionalism, though this often remains invisible to correctionalists, since they identify their own viewpoint with the public good" (396).

2. Designatory: To "enable people to think consciously what they have been only half aware of" and thus "finding the most illuminating language with which to describe people's experiences and actions." Hence "by providing a language which conceptualises the tacit knowledge on which teachers rely, qualitative researchers can aid the development of professional knowledge and skills" (396). This can be a way of rendering explicit forms of good practice.

3. Reflective: Holding up a mirror to the people's experience to see what is going on, rather than what is thought to be going on or wished.

4. Immunological: Hammersley here refers to "the potential for research to immunise us against grandiose schemes of innovation, against raising expectations or setting targets too high; indeed against the 'idolatry of the new' more generally" (398).

5. Corrective: In contrast to point 1, he here refers to the "correction of macro-theoretical perspectives, rather than of the world" (399).

In summary, he takes the line that "qualitative work in particular . . . can remind politicians and policymakers that innovation may have unintended and unforeseen consequences; that what is an improvement is not always a matter of consensus (that there are always diverse perspectives); and that problems often cannot be solved by sheer act of will, by putting in more effort, or through trying to make practices 'transparent'" (400).

Source: Hammersley (2000).

It would require a methods book to develop social work examples of each of these suggested relevance capacities, but by way of single illustration, a good example of the "designatory" capacity of qualitative research can be seen in the twin articles by Neander and Skott (2006, 2008).

Knowing and Practicing Science

One of the umbrellas for this book is a constant effort to problematize and develop a multifaceted argument about the relationship between practice—diversely understood—and science. I am at least as much interested in what practice has to say to research as vice versa. "Knowing" and "doing," research and practice, are not two wholly distinct areas that need mechanisms to connect them but are to a significant degree part and parcel of the other. Two of the most stubborn and difficult-to-avoid options for presenting the relation of science and practice are to give science priority over practice (rationalism) or to give practice priority over science (romantic conservatism). These distinctions are not intended to carry any evaluative judgments—for example, I do not believe that social work research will necessarily be better the closer it is to practice (Shaw and Norton 2007).

We have seen how social workers often face the dominance of science "experts" over practice "beneficiaries," which tends to lead to a deeply unhelpful situation in which practitioners are routinely blamed for their perceived failure to act on the "findings" of research. Kirk and Reid elegantly criticized "practitioner-blame" responses that take the form of discussions of science as progress, of science as having to struggle against "organizational banality," and practitioners as subverting research and easily being threatened. They responded:

> Omitted from these portraits of research is any suggestion that researchers' motives may extend beyond the good and worthy; that scientists are not strangers to aggrandizement or status seeking; that the research process itself can be subjective and biased, sometimes fatally so; or that researchers may have a personal, as well as a professional, stake in persuading practitioners to value their work. There is little recognition that scientific technology has limits or that what researchers have labored to produce may not be particularly usable.
>
> (Kirk and Reid 2002, 190)

One implication of all this is suggested in example 9.3.

EXAMPLE 9.3
Cronbach and Colleagues on Tellable Stories

The social work scientist—and evaluator—face a paradox. "All research strives to reduce reality to a tellable story," but "thorough study of a social problem makes it seem more complicated" (184). Their resolution of this paradox lies in the aphorism that comprehensive examination does not equal exhaustive reporting. "When an avalanche of words and tables descends, everyone it its path dodges" (186).

The main criterion is the extent to which relevant people learn from the evaluator's communications. Therefore the evaluator should seek constant opportunity to communicate with the policy shaping community throughout the research. They believe that "much of the most significant communication is informal, not all of it is deliberate, and some of the largest effects are indirect" (174). Their recommendations are:

1. Be around.
2. Talk briefly and often.
3. Tell stories. Always be prepared with a stock of anecdotes regarding the evaluation.
4. Talk to the manager's sources.
5. Use multiple models of presentation.
6. Provide publicly defensible justifications for any recommended program changes. These will be very different from scientific arguments.

Cronbach is strongly opposed to holding on until all the data is in and conclusions are firm. Influence and precision will be in constant tension, and if in doubt we should always go for influence. Live, informal, quick overviews; responsiveness to questions; the use of film and sound clips; and personal appearances are the stuff of influence. The final report thus acts as an archival document: "The impotence that comes with delay . . . can be a greater cost than the consequences of misjudgement. The political process is accustomed to vigorous advocacy . . . [and] is not going to be swept off its feet by an ill-considered assertion even from an evaluator" (179–180).

Source: Cronbach et al. (1980).

A particularly helpful way in to understanding these issues was provided by the systems theorist Norma Romm. Her basic premise is that "the process of attempting to 'know' about the social world already is an intervention in that world which may come to shape its constitution" (1995, 137). Romm says that "the view that theory is applied in practice and may be tested in that practice, can amount to an unreflected/unreflexive endorsement of a theoretical position" (1996, 25). "One is not just applying 'findings' but intervening in the social discussion in a specific way, that is, in a way which authorises particular conceptions" (1995, 145).

She is not only doubtful about the wisdom of traditional research application perspectives but raises questions about other positions. For example, most but not all participatory approaches advocate reflective deliberation as a means of reaching agreement. Romm chides by saying that such approaches "fail to consider that their 'understanding' *already gears practical activity* in a certain direction, and excludes other options for action" (1995, 151–152). What she means is that participatory models direct research processes toward dialogue, debate, and accommodation of interests. Viewing the world "in a way which suggests that debate can be initiated between participants engaged in purposeful activity, already directs participants towards activities in which they communicate with the hope to achieve an accommodation of interests" (1996, 30) and does not press us to consider the fairness of the accommodation reached or whether contention or confrontation are ever appropriate. Borrowing a phrase from the sociologist Alvin Gouldner, she argues that critical self-reflection must include "acceptance of the need to open oneself to 'bad news' " (1995, 158), which she defines as "news springing from opposing positions."

Approaches that fall within advocacy research recognize this problem and are cautious about pseudocompromises. Yet even here there is a risk that "knowing" predisposes certain action. "By 'knowing' that there 'is' an inevitable and unnegotiable conflict of interests, one may contribute to rendering this so—through the political action that one adopts" (163). This may seem a collection of pessimistic recipes, but I do not think so. Indeed, the abrasive tendencies in the research/practice context are, for me, simply as they should be, nor are they at all special to social work.

The Limits of Social Work Science

In drawing attention to the limits of science in general and social work science in particular, my focus is on what we ought not to expect from science, either because of its inherent characteristics and nature or from the current perimeters of accomplishment. In referring to "current perimeters," this does not assume that the contribution of social work science advances in a progressive way. While the boundaries are fuzzy, to recognize the limits of social work science is not the same as drawing attention to its imperfections and limitations.

We much earlier in this book distanced ourselves from faith in the approaching advent of social work directed and led largely by scientific practice. The advantage of thinking of the limits of social work science is that "acknowledging what our field *does not know* or *cannot do* increases its credibility regarding what it *does know* and *can do*" (Grisso and Vincent 2005, 3). An immediate rejoinder may be that it is commonplace to observe that limits may render a particular method unsuitable for certain purposes yet allow that the method may be ideally suited, reliable, and accurate for other purposes. This is a very weak meaning of "limit" and not one I have in mind. More fundamentally, some may argue along the lines of saying that certain positions—say, "positivism" or "postmodernism"—are inimical with social work research. But it is possible that someone might see certain positions in research as *limits* that are inherent by virtue of the character of research in, for example, the late modern age. Others might see that they are *limitations* that are open—if with difficulty—to transformation. The underlying distinctions here are close to those employed by Collins in his discussion of the various meanings of saying that tacit knowledge is about what we "cannot" know (2010, 88–91).

I continue "thinking in broken images" as to whether social work should be or is a science.[7] Social work should be "scientific" in several ways, and it does and should have something "scientific" to say by way of challenge to other "conventional '-ologies,' each pigeonholed into lesser topics" (Medawar 1984, 3). Yet our deliberations in chapter 7 undergird the thought that "quite ordinary people can be good at science. To say this is not to depreciate science but to appreciate ordinary people" (9). Medawar, however, immediately adds, "But to be good at science one must *want* to be—and must feel a first stirring of that sense of disquiet at lack of comprehension that is one of a scientist's few distinguishing marks" (9), and that entails "power of application and [a] kind of fortitude" (10).

Scientists do not make their discoveries by any one method. There is "no such thing as . . . a schedule of rules by following which we are conducted to a truth" (16). "A scientist commands a dozen different stratagems of inquiry in his approximation to the truth," coupled with "an ability to get things done, abetted by a sanguine expectation of success and that ability to *imagine* what the truth might be" (17–18).

Grasso and Vincent (2005, 3) suggest gains that flow from acknowledging what one's field does or cannot know, and I quote them more fully. " 'Limits' are not necessarily weaknesses:

- Every psychological assessment has limits, and no amount of scientific research can ever create a "limitless" method.
- Limits may render a particular method unsuitable for certain purposes yet allow the method to be ideally suited, reliable, and accurate for other purposes.
- Knowing a method's limits can strengthen the effectiveness and credibility of the method by avoiding its misuse (just as recognizing our own personal limits allow us to focus on our strengths and not make fools of ourselves by trying to excel in areas in which we are unprepared."

Adapting their prescriptions, we might conclude that we should cultivate understanding of what we should be reasonably confident can be done based on the empirical evidence from evidence, understanding, and justice. Practitioners should be aware of what aspects of their work have little or no empirical foundation and what steps they should take in the light of those values and limits.

More broadly, a variety of this kind of question relates to paradigms and incommensurability and to what ways social work science has limits on how far communication is possible between people in different "schools." I endeavored in chapter 2 to make a case for how a mutually understood conversation *is* possible, though one that falls some way short of either shared "beliefs" about the practice of research or a shared understanding of what each "speaker" is saying.

Habermas has talked about forms of ideal speech in which all who participate in discourse must have equal chances to make interpretations, assertions, recommendations, explanations, and corrections. All also must have equal chances to problematize or challenge the validity of these presentations, to make arguments for and against. In ways that have close similarities to aspects of reformist standpoints on the relationship of social work science and justice seen in chapter 8 (for example, in House's unfolding of the terms of a fair evaluation agreement), these two conditions make possible free discourse and pure communicative action where participants by presentative speech acts express equally their attitudes, feelings, and wishes, and also where participants are honest to one another and make their inner nature (intentions) transparent. Reciprocally, all participants must have equal chances to order and resist orders, to promise and refuse, to be accountable for one's conduct and to demand accountability from others.

In doing so social work scientists should not "strain so busily after an obsolete ideal that they neglect the more pertinent aspects of their task" (Madge 1953, 290). They should not want to be "glorifying uncertainty," and "the quest for exactness cannot lightly be renounced." Yet "exact truth is both a proper objective and an unattainable one . . . we have to be content with some residue of uncertainty" (290, 291).

Taking It Further

Reading

Jacob, M. C. 1992. "Science and Politics in the Late Twentieth Century." *Social Research* 59(3): 487–503.

Kirk, S., and W. J. Reid. 2002. *Science and Social Work*. New York: Columbia University Press. Chapter 8.

Walter, I., S. Nutley, J. Percy-Smith, D. McNeish, and S. Frost. 2004. *Improving the Use of Research in Social Care Practice*. London: SCIE. http://www.scie.org.uk/publications/knowledgereviews/kr07.asp.

Kirk and Reid provide a helpful overview of the uses of science, whereas Walter and colleagues develop a more elaborate model of kinds of uses and apply it to social work practice. Jacob is closer to the approach in this chapter.

Bastow, S., P. Dunleavy, and J. Tinkler. 2014. *The Impact of the Social Sciences*. London: Sage.

Merton, R. K. 1971. "Social Problems and Sociological Theory." In *Contemporary Social Problems*, ed. R. Merton and R. Nisbet. New York: Harcourt Brace Jovanovich.

Stokes, D. E. 1997. *Pasteur's Quadrant: Basic Science and Technological Innovation*. Washington, D.C.: Brookings Institution.

Merton's later-career statement of his classic paper, read alongside Stokes's creative and stimulating reworking of the relationship between knowing and doing, are complementary and important texts. Bastow and colleagues provide the best available theoretical and empirical examination of the kneejerk demand that science and research should have "impact."

Task

Carol Weiss (1980) and her colleagues, in the research alluded to in this chapter, interviewed 155 people who held high-level positions in federal, state, and local mental health agencies. Among the questions she asked were:

1. Do you consciously use the results of social science research in reaching decisions on your job?
2. In what ways do you use social science research on your job?
3. Do you seek out research information when you are considering policy or program alternatives?
4. Under what circumstances do you seek research?

The first two questions are about whether the use of research is *conscious* or not. The second two are about *active* information searching.

Arrange to interview a senior official (either known or not known to you), and ask this sequence of questions, modified if necessary, though without changing the nature of the questions. Transcribe and compare the results of the interview with Weiss (1980).

Appendix

Writing Social Work Science

> This writing business, pencils and what-not. Over-rated if you ask me.
>
> —Eeyore (A. A. Milne)

This appendix briefly makes the case that the "voice" as represented and given life in written forms is a central element in what constitutes social work science. It is added at this point because this book has been largely spoken in a conventional scientific voice, mirroring by and large the writing on which it draws.

Timms presented his identity, at least by implication, as being a writer, and in doing so challenged how writing has been seen in social work. Writing in 1969 to the correspondence pages of the final issue of the British social work journal *Case Conference,* he says:

> At some time we ought to try to specify the character and purpose of the social work writer. . . . The writer in this area is still rather a marginal man. He is seen by some as occupied in an essentially second order activity, writing about what others are more properly doing. This is one of the many dichotomies I have come to reject—thought v feeling, theory v practice. In using and studying language a social work writer is laboring at the rock face of the profession.

I consider what may be referred to as a kind of insurgency, on the part of some, in approaches to social work and social science writing. Challenges to writing science came in part from Foucault and his continuous attacks on traditional ways of writing. This is an area where social work has been active, perhaps most frequently in the United States. Stanley Witkin, during the tenure of his editorship of the journal *Social Work* in the opening years of the century, spoke prominently on this theme through his editorials and invited articles. The journal *Qualitative Social Work* has provided a prominent platform for innovative forms of writing. Individual scholars such as Karen Staller have ventured fruitfully into the field. For example, she creatively challenges conventions of journal writing in an article that takes the form of an exchange between author, editors, and reviewers (Staller 2007).

Various points need making by way of elucidation. First, a self-conscious concern with writing social work science is not new. Pauline Young commented on her relationship with the sociologist Robert Park in a memorandum on "Dr. Robert Park as a Teacher": "He insisted that his students carefully examine each sentence and see if 'the words march along.' He would say, '*You* are not writing for professors; train yourself to write for the general public'" (Robert Ezra Park Collection, Box 19, Folder 4). Second, a significant element in more recent developments has been the mutual influence between ethnography and aesthetics (Atkinson 2013). But we should not think this is a question of introducing rhetoric into writing where it previously did not exist. "Style is just as much a matter of choice when the experimentalist writes in a self-conscious, hyper-realistic, attention-grabbing dots-and-dashes fashion—where, for instance, ellipses are used to simulate (and stimulate) the effect of a . . . missed heartbeat—as when the traditionalist falls back on the neutral pale-beige, just-the-facts fashion of scientific reporting" (van Maanen 2011, 5). Third, writing is usually not a choice between one and another kind. Tales of one sort are often nested in tales of another sort. Finally, "once a manuscript is released and goes public . . . the meanings writers may think they have frozen into print may melt before the eyes of active readers," and "different categories of readers will display systematic differences in their perceptions and interpretations of the same writing" (van Maanen 2011, 25).

Why would one write? Motives are not necessarily entirely honorable. Richard Lewontin, the evolutionary biologist, says: "My pleasure in science, I would say, is exactly what every scientist says; my pleasure in science is finding out something that's true. My claim is that most of them are not telling the truth. That most of their pleasure in science comes out not from knowing what's true, but from claiming something that other people think is grand" (Wolpert and Richards 1997, 110). Djerassi comments at the end of his novel *Cantor's Dilemma* that "publications, priorities, the order of authors,[1] the choice of the journal, the collegiality and the brutal competition, the Nobel Prize, *Schadenfreude*—these are the soul and baggage of contemporary science" (1989, 230) and when later interviewed about the work of science said, "it is basically about noble science and Nobel lust" (Wolpert and Richards 1997, 15).

Foucault had a mix of things to say about why he wrote. "I don't write a book so that it will be the final word; I write a book so that other books are possible not necessarily written by me" (O'Farrell 2005, 9). He also said, "everything I have said in the past is of absolutely no importance. You write something when you've already worn it out in your head; drained bloodless thought, you write it and that's that" (11). Yet he also made a distinction between "book experience" (carrying the idea of writing/reading as being transformative and entailing meditation) and "truth book" or "demonstration book"—"namely books which teach something and instruct people on what to think" (112).

Teaching and Writing

The relationship of teaching and research has often been perceived as difficult. We saw in chapter 3 that Kuhn, in his *The Structure of Scientific Revolutions,* had little explicitly to say about teaching, but he seemed to slight it in his remarks about the role of textbooks in rendering scientific revolutions invisible. Critical inferences from similar arguments were explicitly and influentially represented by C. Wright Mills's (1943) famous critique of social pathologists. On textbooks, he talks through how the selection of what to include is not random but exhibits a professional ideology. "The direction is definitely toward particular 'practical problems'—problems of 'everyday life.'" He is critical of how "the ideal of practicality, of not being 'utopian,' operated, in conjunction with other factors, as a polemic against the 'philosophy of history'" (Mills 1943, 168). The "survey" style of textbooks undergirds "an epistemology of gross description" that then lingers in an academic tradition. "The emphasis on fragmentary practical problems tends to atomize social objectives" (168–169) such that "there are few attempts to explain deviations from norms in terms of the norms themselves" (169). As he puts it, "the focus on 'the facts' takes no cognizance of the normative structures within which they lie" (169). He chooses to illustrate this from social work via Mary Richmond's 1917 book *Social Diagnosis*, of which he remarks:

> Present institutions train several types of persons—such as judges and social workers—to think in terms of "situations." Their activities and mental outlook are set within the existent norms of society; in their professional work they tend to have an occupationally trained incapacity to rise above series of "cases." It is in part through such concepts as "situation" and through such methods as "the case approach" that social pathologists have been intellectually tied to social work with its occupational position and political limitations.
>
> (171)

He believes that her book affords "a clue as to why pathologists tend to slip past structure to focus on isolated situations, why there is a tendency for problems to be considered as problems of individuals, and why sequences of situations were not seen as linked into structures" (170), in that by emphasizing "the whole" this assumes there are many parts. This leads to a multicausal approach, which in turn implies that social change will need multiple considerations and not be easy. Whether Mills is correct or not is beside the point,[2] but his argument has value for opening an awareness of how textbooks are far from neutral maps of a given field.

Journals

We considered the role of journals in chapter 9 in relation to arguments regarding the impact of social work science and the risks that more recent developments may militate against true innovation. There are echoes of this in Polanyi's remark:

> Only offerings that are deemed sufficiently plausible are accepted for publication in scientific journals. . . . Such decisions are based on fundamental convictions about the nature of things and about the method which is therefore likely to yield results of scientific merit. . . . These

> beliefs and the art of scientific inquiry based on them . . . are, in the main, tacitly implied in the traditional pursuit of scientific inquiry.
>
> (Polanyi 1966, 64)

The journal paper has almost unquestioned status. Roald Hoffmann, the theoretical chemist, speaks in ways that those in any field can acknowledge:

> Here is the journal report, a product of 200 years of ritual evolution, intended, supposedly, to present the facts and nothing but the facts dispassionately, without emotional involvement, without history, without motivation, just the facts. Well, underneath there's a human being screaming that I'm right and you're wrong. That endows that scientific article with an incredible amount of tension.
>
> (Wolpert and Richards 1997, 24)

I draw on scientists from mainstream sciences to underscore a point made numerous times explicitly and implicitly in this book regarding the "nearness" of much mainstream science to social work science. By failing to see such shared ground we paradoxically fail to recognize the real differences from both the humanities and social sciences. On differences, we noticed in Chapter 2 that Djerassi, who is also a published poet, says "scientific papers are a very ephemeral form of literature. People very rarely reread them. But in poetry . . . the ultimate compliment is to have someone re-read your poems, to remember that book, to remember some metaphor, some nuance that would count for nothing in science" (11). He also has a humanities graduate student as a character in *Cantor's Dilemma* who says, "I don't have anything to 'write up.' With me, I don't really know what I think about a given subject until after I've written" (Djerassi 1989, 65).

Qualitative Science Writing

Qualitative social work and social science writers have taken more interest in the written form than quantitative scholars. Paul Wong, a quantitative scholar at the University of Hong Kong, reflects on writing in example A.1.

EXAMPLE A.1

Paul Wong on Quantitative and Qualitative Writing

I think writing qualitative research papers is much more difficult than writing papers from quantitative research. Writing quantitative papers from observational and experimental studies with good reporting quality can somehow follow a formula or template (for example, CONSORT and STROBE), and the structure and planning of the article take care of themselves (or the journal editors will help you to take care of them!). But writing qualitative papers is a rather creative process. Writing a convincing, objective, scientific story with limited numerical data is a calculative improvisation, much like Jackson Pollock's paintings . . .

He refers in this to CONSORT—Consolidated Standards of Reporting Trials—and STROBE—Strengthening the Reporting of Observational Studies in Epidemiology.[3]

"The ethnographer 'inscribes' social discourse; he writes it down. In so doing he turns it from a passing event, which exists only in its own moment of occurrence, into an account, which exists in its inscriptions and can be reconsulted" (Geertz 1973, 20). In doing so one is trying to "rescue the 'said' of such discourse from its perishing occasions and fix it in perusable terms" (20). Writing is not something that happens as a final account of analysis but happens throughout, through fieldnotes, transcriptions, and memos. "Fieldwork is a site for identity work for the researcher. . . . This identity work is achieved through the textual products of ethnography" (Coffey 1999, 115). Fieldnotes are "the textual place where we, at least privately, acknowledge our presence and conscience." They are "a way of documenting our personal progress" (120). As a consequence, "research . . . includes not only the interaction between the researcher and the issue, but also the interaction between the researchers and their potential readers" (Flick 2006, 406).

What is being eschewed in such developments? There are *genres* of writing—loose sets of criteria for a category of composition. The term is often used to categorize literature and speech, but it is also used for any other form of art or utterance, and it relates to the history of ideas about rhetoric. Classic distinctions of genre in literature are those between prose, drama, and poetry. When applied to academic writing it includes a recognition of ways it is associated with a tradition and a community, with a distinctive style, and forms of expression and vocabulary. The characteristics of this distinctive style when evident in mainstream academic writing include exactness, clear linkages between different aspects of what is written, seriousness of tone, and transparency, for example through the notion of replicability.

Perhaps the most influential characterization of this field is van Maanen's *Tales of the Field* (2011). His distinction between realist and confessional tales has entered the literature, although these are but two of seven distinct genres he identifies. "Of all the ethnographic forms . . . realist tales push most firmly for the authenticity of the cultural representations conveyed by the text" (45). They are marked by authorial invisibility: "a studied neutrality characterizes the realist tale" (47). They also have a documentary style of minute detail, yet in qualitative writing one that may suggest presence. Realist texts convey interpretive omnipotence. But however interpretive authority is achieved, "realist tales are not multivocal texts where an event is given meaning first one way, then another, and then still another. Rather a realist tale offers one meaning and culls its facts carefully to support that reading" (52–53). We quoted the extracts in Example A. 2 in Chapter 3, but we repeat them here for the way they offer an example of such documentary style that also conveys something of authorial presence through extracts from Edith Abbott's housing research in 1930s Chicago.

EXAMPLE A.2

Realist Writing in 1930s Chicago

There are graphic descriptions of the different chosen neighborhoods. For example: "To get a picture of a cross-section of the West Side, it is easy to follow one of the north and south streets and go over the bridge over the river's 'south branch' straight ahead to the bridge over the 'north branch' " (77). Of the Old Lumber Yards district she says: "Dilapidation was everywhere. The cellars, even the first floors, were damp because of the grading up of streets and alleys from three to seven feet above the level of the yards. The walls of the cellars and the floors of the first stories were often decayed and musty, with the water draining down about the foundations. . . . Floors were warped and uncertain, plumbing generally precarious...window panes broken or entirely gone, doors loose and broken, plaster caked and grimy, woodwork splintered and long unvarnished" (80).

It is apparent from this that she sometimes talks as if she is walking the districts, as when she says: "Returning to the busy Halsted Street thoroughfare, one came to . . . " (85), and "Leaving the Lithuanian colony a journey was made around to the back of the great Stock Yards area . . . the area usually referred to as 'back of the yards,' where very congested and insanitary conditions are still to be found." This yields a frequent sense of presence, as in "Looking down the narrow passageways, numerous frame shacks are to be seen on the rear of the lots" (130).

Source: Abbott (1936).

Confessional tales are marked by personalized authority, where "the omnipotent tone of realism gives way to the modest, unassuming style of one struggling to piece together something reasonably coherent out of displays of initial disorder, doubt and difficulty" (van Maanen 2011, 75). "The attitude conveyed is one of tacking back and forth between an insider's passionate perspective and an outsider's dispassionate one" (77). "Confessional accounts . . . can be understood as attempts to deal with problems of trust. . . . If only the writer were able to confess all . . . sufficient information and empathy will be generated to believe his or her story" (Seale 1999, 165). "The implied story line of many a confessional tale is that of a fieldworker and a culture finding each other and, despite some initial spats and misunderstandings, in the end, making a match" (van Maanen 2011, 79). Example A.3 conveys this last point through Elizabeth Whitmore's account of participatory analysis with young people in Toronto.

EXAMPLE A.3

A Confessional Account of Participatory Writing

We first listed all the different ways that we could report what we had learned. These ranged from the usual written report to interactive presentations with the Drop In service users, and seminars and demonstrations to other audiences.

It was totally unrealistic to expect the youth, most of whom had not finished high school, to write a detailed formal report (needed for the agency Board and also for the funders). So how to engage the youth in drafting the written report, if I was to do the actual writing? I had faced this dilemma before . . . and this time, I wanted to devise a process that would result in real ownership of the product by everyone on the team.

Team members began by brainstorming the various parts of the report and then bit by bit, the content—all recorded on flipcharts, of course. I then drafted these into a narrative, which the group then went over (and over and over . . .). One youth, after proclaiming that she did not read and this was boring anyway, pointed out a number of misspelled words and factual errors in one of the drafts—"This is not spelled right. That is wrong!" We did eight drafts of the report and in the end, the youth fully understood and identified with the results. The youth themselves produced as many pieces of the report as possible—cover, index, graphics, charts and tables, appendices. They were deservedly very proud of the final product.

We also did a number of presentations, in which the youth took the lead. We spent quite a bit of time preparing for the first one—a public presentation about our process to the local social work professional association—which got an enthusiastic response. Thereafter, their stage jitters gradually diminished with each one. They presented to the agency board, to the AGM, did a workshop with the wider population of Drop In service users, and were guest speakers at a university seminar.

Perhaps the most interesting reporting mechanism was "The Kit"—a guide for other youth evaluators interested in how to do their own evaluations. "The Kit" was designed and produced entirely by the youth team members, who collectively brainstormed the contents—"what we did," "how we did it" and "tips" (lessons learned)—and then divided up the work. It is the youths' representation of what they learned; its style, content and graphics speak to young people. The result is a boisterous, colourful guide, full of life, energy and humour.

Source: Whitmore (2001).

Not all alternatives to realist writing take the form of experimental writing forms. And some experimental writing forms have a clear realist intent. Over half a century ago C. Wright Mills was arguing for what he called sociological poetry.

> It is a style of experience and expression that reports social facts and at the same time reveals their human meanings. As a reading experience, it stands somewhere between the thick facts and thin meanings of the ordinary sociological monograph and those art forms which in their attempts at meaningful reach do away with the facts, which they consider as anyway merely an excuse for imaginative construction.
>
> (Mills 2008, 34)

Bloor's "Rime of the Globalised Mariner" is an intended use of one manifestation of the form. He borrows the form and meter of Coleridge's "Rime of the Ancient Mariner" to describe and analyze the social situation of the globalized mariner. "The paper aims to be a piece of 'public sociology' and (in seeking to appeal in as vivid a manner as possible) is written in a style that Wright Mills called 'sociological poetry'" (Bloor 2013, 30). "The narrative covers deficiencies in seafarer training, reductions in crew numbers, the consequent long hours and seafarer fatigue, and the failure of global governance of the industry" (30). Example A.4 is an extract from the text.

EXAMPLE A.4

"The Rime of the Globalised Mariner": Sociological Poetry

I had no wish to work on ships—
Filipinos know it's hard—
Mouths were many, jobs were scarce,
From birth my life was marr'd.
From green island homes we travel,
As mariner, nurse, or maid,
And remit to our loved ones
The pittance we get paid.

The Mariner telleth of early hardships and how he and his parents were cheated by the maritime colleges and the crewing agents.

Father scraped up money
For training college fees—
A scam of the local senator,
Whose throat I'd gladly seize.

Filipino maritime training institutions are often controlled by persons with powerful political connections.

> The college had no equipment,
> Just endless, pointless drill,
> No qualifications either—
> The news made my father ill.

The training often follows a military model and is of poor quality. And it does not qualify cadets for certificates of seafarer competency without additional practical experience—"sea time." Most colleges fail to arrange "sea time" for their cadets.

Source: Bloor (2013).

This far from exhausts the range of writing forms that has currency. Poetic forms occur in narrative research, and Stanley Witkin has remarked about "living a constructionist life" (cf. Shaw and Holland 2014, chap. 3). Other instances range from the relatively simple switch to second-person speech that directly addresses someone (see examples 6.3 and 6.4) to writing forms that reflect space and movement in the organization of the text on the page (Martin 2007).

We should not think of writing genres as unchanging. For example, the use of the first-person singular and the active voice have both become more acceptable within mainstream qualitative writing (e.g., Davies 2012). Furthermore, even within a single piece of traditional academic writing there are different forms, for example between the "factual" and the "theoretical." In addition, we should not overdo the distinction between how traditional sciences are done and written about and how the human sciences are done and written about. Interviewed for the BBC, the theoretical chemist Roald Hoffman remarked: "The language of science is incredibly interesting; it's a natural language under strain" (Wolpert and Richards 1997, 24). He said of chemistry, "I love it. I like the subject, its position in between, its compromise between simplicity and complexity. . . . Beauty is in the reality of what's out there, residing at the tense edge where simplicity and complexity contend" (20–21).

There has been a tendency to treat van Maanen's account in romantic Orwellian terms—realist tales bad, confessional tales good. But he insists that "it is important not to judge realist tales too harshly. . . . Realist ethnography has a long and by-and-large worthy pedigree" (54). He warns against denying the "matters covered in a classic realist tale because one prefers a lurid confessional or breezy impressionist tale" and insists "we need more not fewer ways to tell of culture" (140). *All* writing is performative, but "experimental writing cannot be a substitute for a clear sense of *why* one conducts ethnographic fieldwork and *who* one writes for" (Packer 2011, 239). A dose of healthy skepticism is called for in regard to some forms of ethnographic writing, including autoethnography and "so-called ethnographic fiction." "It is too facile, stylistically speaking. It includes far too much emphasis on the feelings and personal experiences of the actual or implied narrator" (Atkinson 2013, 32). Mills was speaking of a particular book, but his words perhaps can be extended slightly when he calls for "the self-discipline of the craftsman of experience" and regrets authors' writing that "often gets in the way of what he would show you" (Mills 2008, 35).

The Author and the Work

Foucault questions the various categories used to organize written material—"author," "work," and "book" (Foucault 1991, 2002). Categorizations are fluid—between an article and a short book, a chapter and an article, a YouTube lecture and an article, and also in how we sequence a person's work. More fundamentally he writes about "the relationship between text and author and with the manner in which the text points to this 'figure' that, at least in appearance, is outside it and antecedes it" (1991, 101).

Academic writing contrives to cancel out the author's individuality and privileged position. But that authorial privilege is preserved by the idea of "the work" of someone. Foucault asks, "what is this curious unity which we designate as a work?" (103). Is it what people write? But if we were to publish, for example, Edith Abbott's work, where would we stop? Would it include the manifold archive of family correspondence in the University of Chicago's Special Collections? Maybe not, but why not? But what about memos drafted to develop the Graduate School of Social Service Administration, also in the Special Collections? Then where do we place the transcribed interviews with the good and the great garnered through the NASW oral history project in the 1970s and, for the United Kingdom, the interviews and transcripts deposited at the Modern Records Centre?[4] A comparison of an average academic social work curriculum vitae from the United States with one from the United Kingdom would show how the boundaries of what counts as "work" are broader in the former. Finally: "How can one define a work amid the millions of traces left by someone after [their] death?" (104). "The word *work* and the unity it designates are probably as problematic as the status of the author's individuality" (104).

If we are to grasp the significance of science writing, we need, among other aspects, to consider the meaning and problems associated with the use of the author's name—"It has other than indicative functions: more than an indication, a gesture, a finger pointing at someone, it is the equivalent of a description" (105). An author's name "performs a certain role with regard to narrative discourse, assuming a classificatory function. Such a name permits one to group together a certain number of texts, define them, differentiate them from and contrast them to others. In addition it establishes a relationship among the texts" (107). I encountered something like this at the proposal review stage for this book. An anonymous reviewer suspected—though could find no definite evidence in the proposal—that I would write what may amount to an antipositivist tract, saying, for example, "I cannot be sure from the prospectus that the approach will be as antipositivist as I suspect it might be. . . . If the book takes the epistemological approach that I suspect it will take . . . " This suspicion probably stemmed from a certain reading of "Shaw" as a descriptive designation. To take a social work example, the name "Reid," understood of William J. Reid, "manifests the appearance of a certain discursive set and indicates the status of this discourse within a society and a culture" (Foucault 1991, 107), restricting "society" and "culture" in this context to American social work science. The name Reid hence has a different kind of author function than that of, say, Shakespeare. If we think of Reid again, as lead author of *Task-Centered Casework*, the name forms part of a discourse. It entails, for example, a kind of ownership, with associated rights. The identity of an "author" is constructed over time, and in this sense "the author does not precede the works" (118–119). To quote Foucault once more: "He who writes does not have the right to give orders as to the use of his writings" (O'Farrell 2005, 55). In Reid's case we are not thinking of him only, or even perhaps primarily, as the author of one or another book but, as we saw in chapter 5, as the inventor of something called task-centered social work. The point is more apparent if we think of Freud or Marx, both of whom have established the possibility—almost without limit—of discourse. "They have created a possibility for something other than their discourse, yet something belonging to what they founded" (Foucault 1991, 114).[5]

Reading

Djerassi, C. 1989. *Cantor's Dilemma*. New York: Penguin.
Staller, K. 2007. "Metalogue as Methodology." *Qualitative Social Work* 6(2): 137–157.
van Maanen, J. 2011. *Tales of the Field: On Writing Ethnography*. Chicago: University of Chicago Press.

The possible directions for pursuing this further are clear from what I have said. I would recommend to read an example of writing. Staller has exemplified this as much as anyone, but I would also recommend having fun by reading Carl Djerassi's novel. Van Maanen is the standard contribution in this field and is relevant beyond ethnography.

Notes

Introduction

1. *Passionate Minds* (Wolpert and Richards 1997) is the title of a book that I will draw on in various places.
2. I use "statement" here in a Foucauldian sense, as "particular modalities of existence" (Foucault 2002, 21) and a way of speaking that can be described and has sets of rules for how to speak. "One can define the general set of rules that govern the status of these statements, the way in which they are institutionalised, received, used, re-used, combined together, the mode according to which they become of objects of appropriation, instruments of desire or interest, elements for a strategy" (129).
3. I am indebted to Riesman and Becker for this metaphor (1984, xii).
4. Jonathan Franzen, "The Path to Freedom," *Guardian* (May 26, 2012).

1. Talking Social Work Science

1. To trace a line of appreciation from Hill to Richmond to Burgess is to strike a rarely heard note about the relationship between the United States and the United Kingdom, and more so of the relationship between sociology and

social work. Burgess is quoting Hill to make a point. "Existing case records seldom, or never, picture people in the language of Octavia Hill, with their 'passions, hopes, and history' or their 'temptations,' or 'the little scheme they have made of their lives, or would make if they had encouragement.' The characters in case records do not move, and act, and have their being as persons. They are depersonalized, they become Robots, or mere cases undifferentiated except by the recurring problems they present" (Burgess 1928, 526–527).

2. For example, the word "case" well rewards determined pursuit in this connection (Shaw forthcoming).
3. "Object" and "subject" once made the distinction we are familiar with today but the opposite way round (Williams 1983).
4. E.g., the relation of word as signs, their relation to concepts and so on.
5. However, it does appear to be a likely response from some. Fischer (1993), for example, anticipated the soon-approaching day when social work would be wholly formed by empirical practice.
6. Even the half-alert reader will detect that I do not discuss the place of postmodernism. This is not from an Anglo-Saxon disdain for Continental philosophy. Rather, postmodernist rhetoric has been taken up in social work more in the language of practice than research. While theoretically important, I concur with Matthewman and Hoey's remark that "it is almost as if postmodernism never occurred" (2006, 530). I also agree that "many of the questions posed by postmodernism will rise again . . . agency versus structure, Enlightenment versus Romanticism, humanism versus science, relativism versus realism, who can speak and what can be said" (542).
7. Though Paul Helm (personal communication, January 2015) commented that "Aristotle was an essentialist of sorts."
8. I am grateful to an anonymous reviewer for pointing to the work of Heather Douglas, where this is defined as meaning only scientific values are permitted, and for clarifying that when such positions are claimed it is not clear if the accomplishment of value-freedom is done by the individual or the group, e.g., by anonymous peer review (Douglas 2009).
9. MacKay, a major figure in the development of communication and brain science, had no time for equivocation. "The central point for the Christian . . . is that even where *he* may find it difficult or impossible to arrive at a value-free description of a human situation, he is under the judgement of One who *knows* the way things are, for it is He who creates them and now holds them in being, just as they are. . . . The ideal of value-free knowledge is the representation of what the Creator has provided for me to reckon with, as it is, whether I like it or not" (1981, 16). He goes on: "Christians who believe

that objectivity is a duty to the Creator, before whom the scientist is under judgment, have no need of the scientistic *hubris* of Positivism to back up their emphasis" (17).

10. This has connections with the idea of truth as justified true belief (cf. Shaw 2012c).
11. Durkheim's sociology has received a more positive reappraisal in recent years. An anonymous reviewer usefully observed that "Durkheim's sociology, for all its positivism, assumes/requires some notion of the social whole (e.g., religion is a collective phenomenon not an individual one)."
12. Campbell thought his article (1964) on distinguishing differences of perception from failures of communication as "my own best hermeneutic achievement" (Campbell 1991, 594). Campbell's 1991 paper is a profound and measured defense of a complex position.
13. Florian Znaniecki (1882–1958). Most famous as coauthor, with W. I. Thomas, of *The Polish Peasant*. There is an impressive and comprehensive essay on Znaniecki at http://en.wikipedia.org/wiki/Florian_Znaniecki.
14. For example, in her *Grounded Theory, Analytic Induction, and Social Work* e-book. https://read.amazon.co.uk/?asin=B0030EFX82, and in Gilgun (2005).
15. For a discussion of Becker's subsequent thinking, see Hammersley (2008, 80–84).
16. The role of interests is dealt with fully elsewhere, most extensively in chapters 5 and 8.
17. It is curious that no such critique seems to have been attempted.

2. Doing Social Work Science

1. E.g., Timms's work on art and science in his *Language of Social Casework* (1968).
2. This is probably derived from Polanyi (1966).
3. I quote this translation with the permission of Taka Asano, who at the time was completing a Ph.D. on postqualifying formal and informal learning among Japanese social workers.
4. There is a classic philosophical problem here first set out by Plato as Meno's paradox. Polanyi restates the problem: "if all knowledge is explicit, i.e. capable of being clearly stated, then we cannot know a problem or look for its solution." But if we nonetheless do solve problems "we can know things, and important things, that we cannot tell" (1966, 22). Hence "we can have a tacit knowledge of as yet undiscovered things," and "since we have no explicit knowledge of these unknown things, there can be no explicit justification of a scientific truth" (23).

5. It is also a perennial challenge for efforts at reflexivity in qualitative research, although the almost complete insensitivity to the problem in the majority of quantitative social work science is a still deeper problem.
6. I have explored the possibilities of arts-based approaches as mediating research and practice in Shaw (2011a).
7. The antiessentialist mood in Western academic communities might favor a broader interest in "spirituality," but that field of writing is less pointed in its challenge, as is work that applies standards of evidence-based practice to agencies with a faith-based constitution. I could, however, have opted for the writing of Catholic social work scholars such as Noel Timms.
8. University of Chicago Office of the President, Box 77, Folder 9: "Small, Albion, W. 1992–1924."
9. In these interviews Neumeyer, Johnson, Carter, Cottrell, and Haynor all refer to ways in which sociology both drew on (almost always) men who came from family backgrounds that included ministers and how sociology increasingly replaced that identity.
10. An important contribution to this debate has been made by those who argue that the writing of the fourteenth-century (Christian calendar) writer Ibn Khaldun provides the basis for a truly indigenous Arab-Muslim sociology, through his organizing concept of *Al'Assabiyya*, through which he sought to explain how and why things are as they are (Dhaouadi 1990).
11. Recognizing that Enlightenment skepticism continues in late-modern social work science, we can note how Francis Bacon exemplified the freedom from Greek rationalism. Hooykaas remarked how he "analysed what had been the fault of the past, and why science had born no fruit up till then. 'We will have it that all things are as in our folly we think they should be, not as seems fittest to divine wisdom, not as they are found to be in fact. . . . We desire to be like God and follow the dictates of our own reason' " (quoted in Hooykaas 1960, 15). Hooykaas concludes elsewhere that "the eighteenth century, which is the first wholly modern century, continued in many respects the traditions of scholarship established on biblical grounds; but its spirit was rationalistic in essence, although empiricist by compromise" (1957, 22). "The essential difference between the eighteenth-century Enlightenment and the Puritan Enlightenment is that, to the Puritans it was not *freedom* which led to Truth, but it was truth which led to freedom" (23). Or "we could perhaps better say that to the Puritans it was Truth which led to the freedom necessary to find truth" (23).
12. Earlier in his career he would have said quite different things. In the 1970s he went as far as to say, "Knowledge can only be a violation of the things to be known, and not a perception, a recognition, an identification of or with those things" (O'Farrell 2005, 67).
13. Quoted in Wolpert et al. (1997, 14).

14. The eventual article was published as E. J. Mullen, "Reconsidering the 'Idea' of Evidence in Evidence-Based Policy and Practice," published online (March 20, 2015), DOI: 10.1080/13691457.2015.1022716.
15. Interviewed for *Times Higher Education* (July 22, 2010).

3. Historical Moments for Social Work and Science

1 Quoted in Wolpert and Richards (1997, 154).
2. Who actually invented paper is immaterial. The point is that almost every Chinese schoolchild and almost no American schoolchildren will know the story of Ts'ai Lun, an official of the Imperial Court. What "everyone knows" will usually prove a precarious assumption.
3. http://historyofsocialwork.org/eng/index.php.
4. Digitized at http://www20.us.archive.org/stream/longviewpapersad00joan/longviewpapersad00joan_djvu.txt. I am indebted to Irene Lewin for drawing this paper to my attention. But see Seed (1973) for an analysis of social work in Britain as caught up in different social movements.
5 Eileen Younghusband, WiseARCHIVE, the Cohen Interviews, http://www.wisearchive.co.uk/home/.
6. Margaret Simey, speaking around 1980, remarks of old age that "I think that people like Barbara Wootton are unhappy because they haven't found a principle that takes them through their sad days and their difficult days" (Simey, WiseARCHIVE, the Cohen Interviews, http://www.wisearchive.co.uk/home/). The best study of Wootton's role is Oakley (2011).
7. Personal communication (August 21, 2014). Davies's position can be connected in this respect to that of Perlman when she grumbled about "the caseworker's aspiration (rarely his client's) to goals that are illusory: achieving 'cure' for instance, rather than the more realistic goal of achieving some restored or new equilibrium" (in her foreword to Reid and Shyne 1969, vi).
8. Burawoy argued that "in its beginning sociology aspired to be . . . an angel of history, searching for order in the broken fragments of modernity, seeking to salvage the promise of progress" (2005, 5).
9. I am indebted to Mirja Satka for this observation.
10. Judgments of what counts as sentimental, romantic, and so on are liable to the same sentimentality. Pickering and Keightley, for example, suggest how the meanings of nostalgia are multiple "and so should be seen as accommodating progressive as well as regressive stances and melancholic attitudes" (2006, 919).
11. http://www.historyofsocialwork.org/eng/index.php.
12. http://www.kcl.ac.uk/sspp/policy-institute/scwru/swhn/index.aspx.
13. I present the evidence and discussion fully in Shaw (2014b).

14. Burrow is actually talking of the writing of John Stuart Mill's father, James Mill.
15. http://www.albany.edu/president/awards/1998/reid.html.
16. My argument at this point is open to challenge on the grounds that it may not be plausible to think of social work science as accumulating anomalies in the sense Kuhn intended. This is because an anomaly presumes the existence of "normal science," which in turn assumes the idea of shared paradigmatic presuppositions that, as they face anomalies, lead in time to scientific revolutions. Given that Kuhn never seems to argue that the social sciences are conducted within one or more paradigms, it seems to raise the question for social work whether there are "anomalies," or crises, or revolutions, or "fundamental novelties." However, this in no way precludes the occurrence of controversies, passionately fought on all sides.

4. Technology and Social Work

1. http://www.theguardian.com/science/political-science/2013/sep/19/harold-wilson-white-heat-technology-speech. Delivered October 1, 1963, to a Labour Party Conference.
2. May 24, 2014, on "Who's Really Driving Change?"
3. This argument directly parallels ways in which a linear relationship between research and practice may be reformed (Shaw 2012).
4. http://www.bbc.co.uk/news/science-environment-17560379. "The head of an experiment that appeared to show subatomic particles travelling faster than the speed of light has resigned from his post. Prof Antonio Ereditato oversaw results that appeared to challenge Einstein's theory that nothing could travel faster than the speed of light. Reports said some members of his group, called Opera, had wanted him to resign. Earlier in March, a repeat experiment found that the particles, known as neutrinos, did not exceed light speed."
5. I am indebted to an anonymous reviewer for drawing this very helpful connection to my attention.
6. Names here and in other transcript extracts through the book are anonymized.
7. Unpublished data from a study by Shaw and Lunt.
8 The website from which this text is taken is no longer available.

5. The Social Work Science Community: Controversies and Cooperation

1. "Crude" because multiple names on a given article may not indicate joint writing, although it does seem to imply joint research.

2. I took a string of eighty recent articles from each of *Social Work Research* and *The British Journal of Social Work*. I omitted editorial essays.
3. Shimuzu and Hirao (2009) model the kind of serious empirical inquiry that is needed. They found greater development of networks in the United States than elsewhere, and high mobility was a factor for development of networks. This interesting question of differences in networks between the United States and Europe raises for me the wider question of similarities and differences between universities in Europe and the United States. Fuller (2009) discusses this issue.
4. The study by Ian Shaw, Neil Lunt, and Hannah Jobling is as yet unpublished.
5. This statement was taken from a UK government website, since closed, at http://www.everychildmatters.gov.uk/strategy/childrensfund/ (accessed July 17, 2008). Information is available at http://www.ofsted.gov.uk/resources/childrens-fund-first-wave-partnerships.
6. Indeed, not only about social work but with social workers, such as Stuart Queen (e.g., Queen 1928).
7. September 25, 2014. The figures refer to 2013.
8. http://www.albany.edu/president/awards/1998/reid.html. "Inventor" is not being used with the elaborated sense developed here.
9. Personal communication, March 31, 2012.
10. The Community Service Society of New York—the agency in which the "Casework Methods Project" took place. The project design was first reported in Shyne (1965). Findings were reported Reid and Shyne (1969).
11. Personal communication, March 10, 2014.
12. On the wider context, see Epstein (1995).
13. There may be some basis for Fortune's assertion. The phrase "empirical practice," for example, occurs only six times in *The British Journal of Social Work* from 1996, twice by the author. The expression "evidence-based practice" occurs 271 times.
14. The term "strong" is associated with the position of the "Strong School" of Edinburgh sociology.

6. Social Work Science and Evidence

1. I treat "evidence-based practice," "scientific practice," and "empirical practice" interchangeably in this chapter. I am aware that this is a position accepted by some but not by others, who would argue in a U.S. context that empirical practice is a U.S. strand of work while evidence-based practice is a distinct import from the United Kingdom. It is also the case that the phrases "empirical practice" and "scientific practice" are rarely present in discussions outside

the United States. I reflect that debate in chapter 5 and give relevant and striking figures in the endnotes to that chapter, but here I am concerned more broadly with the role of evidence in social work science.

2. This does not apply to critical realism as developed in relation to social work evidence by Pawson and others (e.g., Pawson et al. 2011).
3. Madge, it should be noted, is not dismissing such approaches and is happy to accept this as an inevitable principle of science.
4. I have traced this in Shaw and Holland (2014, 47–51).
5. As of December 2014, chair of the National Trust in the United Kingdom.
6. *The Guardian* is the main left-leaning liberal newspaper in the United Kingdom. The remark was made on the *BBC Morning Service* on July 12, 1999. On Simon Jenkins, see http://www.theguardian.com/profile/simonjenkins.
7. Nonetheless, as soon as someone argues that knowledge of, for example, service outcomes is the most important kind of knowledge, a semiabsolutist hierarchy is smuggled back in.
8. Roughly speaking, they are either regarded as inevitable sources of bias that need to be controlled in ways to which we referred early in the chapter or as inevitable elements that underscore the need for discretionary judgment.
9. Yet in an earlier book Madge more wisely concluded, "if we are honest we have to admit that the first century of social science has left us somewhere short of victory" (Madge 1953, 290). A more sophisticated view of progress can be seen in the work of Donald Campbell and those associated with his position, in their cautious albeit committed acceptance of a social evolutionary model for social science (Brewer and Collins 1981).
10. There is a significant methodological issue entailed here, in that enumerative analyses of causal elements are unlikely to answer such questions. Ragin has argued for a reinstatement of methods of case-based analytic induction and for qualitative comparative analysis (cf. Ragin 1989, 1997). Byrne (2011) also carries very good discussions of causality despite his rather tendentious style.
11. The sense of professionals mounting the battlements can be seen from the dates. The book came out in December 1965, the review was submitted before mid-April, and it was published by June 1966.

7. Social Work Science and Understanding

1. This was originally located at www.topssengland.net.
2. In an interview by Karyn Cooper and Naomi Hughes: http://cooperwhite.com/geertzvideo.html. See also Cooper and Hughes (2015, 30).

3. Schutz routinely uses language that we read as gendered.
4. Relevances are of different levels of intensity, and it is "the world within my reach"—either actual or potentially so—which comprises "the core of primary relevance" (Schutz 1967b, 127). Our relevances are either intrinsic—the outcomes of our chosen interests—or imposed, for example, by bereavement, destiny, or the feeling of being "thrown into the world" (Heidegger). For myself, if I ask someone to get me a ticket for a Bob Dylan concert, then my relevance is intrinsic, but theirs is imposed.
5. Giddens is here arguing from logic, but he also says the distinction is also partly empirical, in that social science studies "institutional reflexivity."
6. In fairness to Polanyi he does immediately say "this fact seems obvious enough; but it is not easy to say exactly what it means" (Polanyi, 1966:4).
7. As an aside, this is why qualitative methods are thus much more likely to be dealing with tacit knowledge than structured and distanced methods. In this particular sense it is plausible to speak of qualitative and quantitative methods as having aspects of incommensurability.
8. For an interesting example of a research method developed in part to make such knowledge explicit, see Rodriguez and Ryave (2002). For a suggested application to social work practice, see Shaw (2011a, 100–102).
9. http://plato.stanford.edu/entries/peirce/#dia.
10. This is not, of course, to say they are the same. Hoffmann subsequently adds: "One thing that's different is . . . that science is infinitely paraphraseable and art is not" (Wolpert and Richards 1997, 23).

8. Social Work Science and Justice

1. See, for example, http://www.socialworkers.org/profession/centennial/milestones_3.htm.
2. Cronbach and his collaborators throughout use the gendered pronoun.
3. Gergen nonetheless says that "empirical findings can vitalise discussion of moral and political issues" and refers to Milgram's classic laboratory experiments on obedience (61–62). (Interesting alternative interpretations of Milgram's work have come under more recent attention, suggesting that it was the perceived standing of "science" that explains the willingness of participants to seem willing to inflict pain on hidden research subjects.)
4. She is not a naïve rationalist. She adds the rider, referring to her father, that this "is not to say that evidence plus opinion, in the kind of clever networking Richard Titmuss and his disciples were so good at . . . isn't the most effective option of all" (259).

9. Impacts and Influences

1. This report has subsequently been archived.
2. Oancea and Furlong (2007), in a significant article, take exception to parts of Stokes's argument.
3. http://www.rcuk.ac.uk/kei/impacts/Pages/home.aspx.
4. http://blogs.lse.ac.uk/impactofsocialsciences/2013/07/08/engaging-people-in-making-history/?pfstyle=wp.
5. For any given year, the number of times articles published in that year have cited articles in the journal in the previous two years is counted. Suppose there were 15 and 10 citations respectively, making 25. Then the number of articles published in that journal for those same years also is counted. Let us suppose this was 50 articles. The first number is divided by the second number, which gives us 0.5.
6. In a lecture to the University of York in 2014.
7. The reference is to Robert Graves's well-known "In Broken Images."

Appendix. Writing Social Work Science

1. A character later refers to "that most ignominious fate of a scientific collaborator, the most anonymous of appellations: *et al.*" (83).
2. Incidentally, Mills's understanding of the significance of "situation" in social work of that era is open to serious challenge (Shaw forthcoming).
3. http://www.consort-statement.org/; http://www.strobe-statement.org/.
4. http://www2.warwick.ac.uk/services/library/mrc/explorefurther/speakingarchives/socialwork/.
5. This process can work temporally in reverse. What we think of as the late-modern period has gained much of its discursive identity through the development of something we call "postmodernism."

References

Archives

Edith and Grace Abbott Papers. University of Chicago Special Collections.
Ernest Burgess Papers. University of Chicago. Special Collections.
George Herbert Mead Papers. University of Chicago. Special Collections.
Interviews with Graduate Students of the University of Chicago. Department of Sociology, University of Chicago. University of Chicago Special Collections.
Robert Ezra Park Collection. Special Collections Research Center, University of Chicago Library.
Social Science Research Committee (SSRC) Department of Economics. Records, 1912–1961. University of Chicago. Special Collections.
University of Chicago, Office of the President. University of Chicago Special Collections.
WiseARCHIVE. The Cohen Interviews. http://www.wisearchive.co.uk/home/.

Books and Articles

Abbott, A. 1988. *The System of Professions: An Essay on the Division of Expert Labor.* Chicago: University of Chicago Press.

Abbott, A. 1995. "Boundaries of Social Work or Social Work of Boundaries?" *Social Service Review* 69(6): 546–562.

Abbott, E. A., 1931. *Social Welfare and Professional Education*. Chicago: University of Chicago Press.

Abbott, E. A. 1936. *The Tenements of Chicago, 1908–1935*. Chicago: University of Chicago Press.

Abrams, P. 1984. "Evaluating Soft Findings: Some Problems of Measuring Informal Care." *Research, Policy, and Planning* 2(2): 1–8.

Ajzenstadt, M., and J. Gal. Forthcoming. "Social Work and the Construction of Poverty in Palestine in the 1930s." *Qualitative Social Work*. DOI: 10.1177/1473325014527489.

Atkinson, P. 2013. "Ethnographic Writing, the Avant-Garde, and a Failure of Nerve." *International Review of Qualitative Research* 6(1): 19–35.

Atkinson, P., A. Coffey, S. Delamont, J. Lofland, and L. Lofland. 2001. *Handbook of Ethnography*. London: Sage.

Atkinson, P., S. Delamont, and M. Hammersley. 1988. "Qualitative Research Traditions: A British Response to Jacob." *Review of Educational Research* 58(2): 231–250.

Attlee, C. R. 1920. *The Social Worker*. The Social Service Library. London: G. Bell and Sons. http://archive.org/details/socialworker00attliala.

Babbage, C. 1830. *Reflections on the Decline of Science in England and on Some of Its Causes*. London: B. Fellowes.

Bastow, S., P. Dunleavy, and J. Tinkler. 2014. *The Impact of the Social Sciences*. London: Sage.

Becker, H. 1960/1970. "Notes on the Concept of Commitment." In *Sociological Work*. New Brunswick, N.J.: Transaction.

Becker, H. 1963a. "Becoming a Marijuana User." In *Outsiders: Studies in the Sociology of Deviance*. Newbury Park, Calif.: Sage.

Becker, H. 1963b. *Outsiders: Studies in the Sociology of Deviance*. New York: Free Press.

Becker, H. 1970. *Sociological Work*. New Brunswick, N.J.: Transaction.

Becker, H. 1993. "Theory: The Necessary Evil." In *Theory and Concepts in Qualitative Research: Perspectives from the Field*, ed. D. Flinders and G. Mills. New York: Columbia University Press.

Becker, H. 1999. "The Chicago School, So-Called." *Qualitative Sociology* 22(1): 3–12.

Becker, H., and J. Carper. 1956/1970. "The Development of Identification with an Occupation." In *Sociological Work*, by H. Becker. New Brunswick, N.J.: Transaction.

Becker, H., and J. Carper. 1957/1970. "Adjustment of Conflicting Expectations in the Development of Identification with an Occupation." In *Sociological Work*, by H. Becker. New Brunswick, N.J.: Transaction.

Beresford, P. 2010. "Re-examining Relationships Between Experience, Knowledge, Ideas, and Research: A Key Role for Recipients of State Welfare and Their Movements." *Social Work & Society* 8(1): 6–21. http://www.socwork.net/sws/article/view/19.

Billig, M. 2004. "Methodology and Scholarship in Understanding Ideological Explanation." In *Social Research Methods*, ed. C. Seale. London: Routledge.

Bloor M. 1978. "On the Analysis of Observational Data: A Discussion of the Worth and Uses of Inductive Techniques and Respondent Validation." *Sociology* 12: 545–552. Reprinted in B. Smart et al., eds. 2013. *Observational Methods*. London: Sage.

Bloor, M. 1997. "Addressing Social Problems Through Qualitative Research." In *Qualitative Research: Theory, Method, and Practice*, ed. D. Silverman. London: Sage.

Bloor, M. 2001. "On the Consulting Room Couch with Citizen Science: A Consideration of the Sociology of Scientific Knowledge Perspective on Practitioner-Patient Relationships." *Medical Sociology News* 27(3): 19–40.

Bloor, M. 2010. "Commentary: The Researcher's Obligation to Bring About Good." *Qualitative Social Work* 9(1): 17–20.

Bloor, M. 2013. "The Rime of the Globalised Mariner: In Six Parts (with bonus tracks from a chorus of Greek shippers)." *Sociology* 47(1): 30–50.

Bloor, M., and N. McKeganey. 1989. "Ethnography Addressing the Practitioner." In *The Politics of Field Research: Sociology Beyond Enlightenment*, ed. J. Gubrium and D. Silverman Newbury Park, Calif.: Sage.

Bloor, M., and F. Wood. 2006. *Keywords in Qualitative Methods: A Vocabulary of Research Concepts*. London: Sage.

Blumer, H. 1969. *Symbolic Interactionism: Perspective and Method*. Englewood Cliffs, N.J.: Prentice-Hall.

Blyth, E., S. M. Shardlow, H. Masson, K. Lyons, I. Shaw, and S. White. 2010. "Measuring the Quality of Peer-Reviewed Publications in Social Work: Impact Factors—Liberation or Liability?" *Social Work Education* 29(2): 120–136.

Bodner, G. M. 1991. "Ethics in Science." *CHEMTECH* 21: 274–280.

Bogdan, R., and S. Taylor. 1994. "A Positive Approach to Qualitative Evaluation and Policy Research in Social Work." In *Qualitative Research in Social Work*, ed. E. Sherman and W. Reid. New York: Columbia University Press.

Boghossian, P. 2006. *Fear of Knowledge*. New York: Oxford University Press.

Bonefeld, W. 2014. *Critical Theory and the Critique of Political Economy*. New York: Bloomsbury.

Brante, T. 2010. "Professional Fields and Truth Regimes: In Search of Alternative Approaches." *Comparative Sociology* 9: 843–886.

Brante, T. 2011. "Professions as Science-Based Occupation." *Professions and Professionalism* 1(1): 4–20.

Brante, T., and A. Elzinga. 1990. "Towards a Theory of Scientific Controversies." *Science Studies* 2: 33–46.

Breckinridge, S. 1924/1932. *Family Welfare Work in a Metropolitan Community: Selected Case Records*. Chicago: University of Chicago Press.

Brewer, M. B., and B. E. Collins. 1981. "Perspectives on Knowing: Six Themes from Donald T. Campbell." In *Scientific Inquiry and the Social Sciences*, ed. M. Brewer and B. Collins. San Francisco: Jossey Bass.

Brodie, I. 2000. "Theory Generation and Qualitative Research: School Exclusion and Children Looked After." Paper from ESRC seminar series "Theorising Social Work Research." Unpublished.

Brown, M. B. 2015. "Politicizing Science: Conceptions of Politics in Science and Technology Studies.' *Social Studies of Science* 45 (1): 3–30

Bull, R., and I. Shaw. 1992. "Constructing Causal Accounts in Social Work." *Sociology* 26(4): 635–649.

Bulmer, M. 1986. *Neighbours: The Work of Philip Abrams*. Cambridge: Cambridge University Press.

Burawoy, M. 2005. "For Public Sociology." *American Sociological Review* 70(1): 4–28.

Burgess, E. W. 1927. "The Contribution of Sociology to Family Social Work." *The Family* (October): 191–193.

Burgess, E. W. 1928. "What Social Case Records Should Contain to Be Useful for Sociological Interpretation." *Social Forces* 6(4): 524–532.

Burrow, J. W. 1970. *Evolution and Society*. Cambridge: Cambridge University Press.

Byrne, D. 2011. *Applying Social Science*. Bristol: Policy.

Campbell, D. T. 1964. "Distinguishing Differences of Perception from Failures of Communication in Cross-Cultural Studies." In *Cross-Cultural Understanding: Epistemology in Anthropology*, ed. F. Northrop and H. Livingston. New York: Harper and Row.

Campbell, D. T. 1969. "Reforms as Experiments." *American Psychology* 24(4): 409–429.

Campbell, D. T. 1979. "Degrees of Freedom and the Case Study." In *Qualitative and Quantitative Methods in Evaluation Research*, ed. T. D. Cook and C. S. Reichardt. Beverly Hills, Calif.: Sage.

Campbell, D. T. 1981. "Comment: Another Perspective on a Scholarly Career." In "Perspectives on Knowing: Six Themes from Donald T. Campbell," in *Scientific Inquiry and the Social Sciences*, ed. M. Brewer and B. Collins. San Francisco: Jossey Bass.

Campbell, D. T. 1991 "Coherentist Empiricism, Hermeneutics, and the Commensurability of Paradigms." *International Journal of Educational Research* 15(6): 587–597.

Carey, J. 1975. *Sociology and Public Affairs: The Chicago School*. Beverly Hills, Calif.: Sage.

Carey, M. 2014. "Mind the Gaps: Understanding the Rise and Implications of Different Types of Cynicism Among Statutory Social Workers." *British Journal of Social Work* 44(1): 127–144.

Chambon, A. 2008. "Social Work and the Arts: Critical Imagination." In *Handbook of the Arts in Qualitative Research: Perspectives, Methodologies, Examples*, ed. J. G. Knowles & A. Cole. Thousand Oaks, Calif.: Sage.

Chambon, A. 2012. "Disciplinary Borders and Borrowings: Social Work Knowledge and Its Social Reach: A Historical Perspective" *Social Work and Society* 10(2). http://www.socwork.net/sws/article/view/348.

Chapin, F. S. 1920. *Field Work and Social Research*. New York: Appleton-Century.

Chapin, F. S. 1936. "Social Theory and Social Action." *American Sociological Review* 1(1): 1–11.

Chapin, F. S. 1947. *Experimental Designs in Sociological Research*. New York: Harper.

Chapin, F. S., and S. A. Queen. 1937/1972. *Research Memorandum on Social Work in the Depression*. Social Science Research Council Bulletin 39. New York: Arno.

Checkland, P. 1999. *Soft Systems Methodology in Action*. Chichester: John Wiley and Sons.

Checkland, P., and S. Holwell. 1997. *Information, Systems, and Information Systems: Making Sense of the Field*. Chichester: John Wiley and Sons.

Checkland, P., and S. Holwell. 1998. "Action Research: Its Nature and Validity." *Systemic Practice and Action Research* 11(1): 9–21.

Chelimsky, E. 1997. "Thoughts for a New Evaluation Society." *Evaluation* 3(1): 97–118.

Cicourel, A. V. 1985. "Text and Discourse." *Annual Review of Anthropology* 14: 159–185.

Coffey, A. 1999. *The Ethnographic Self: Fieldwork and the Representation of Identity*. London: Sage.

Collins, H. 2010. *Tacit and Expert Knowledge*. Chicago: University of Chicago Press.

Collins, H. 2014. "Rejecting Knowledge Claims Inside and Outside Science." *Social Studies of Science* 44(5): 722–735.

Collins, H., and R. Evans. 2002. "The Third Wave of Science Studies: Studies of Expertise and Experience." *Social Studies of Science* 32(2): 235–296.

Collins, H., and R. Evans. 2007. *Rethinking Expertise*. Chicago: University of Chicago Press.

Collins, H., and T. Pinch. 1988. *The Golem: What You Should Know About Science*. Cambridge: Cambridge University Press.

Collins, H., and T. Pinch. 1998. *The Golem at Large: What You Should Know About Technology*. Cambridge: Cambridge University Press.

Comstock, D. 1982. "A Method for Critical Research." In *Knowledge and Values in Educational Research*, ed. F. Bredo and W. Feinburg. Philadelphia: Temple University Press.

Evans, R., and A. Plows. 2007. "Listening Without Prejudice? Rediscovering the Value of the Disinterested Citizen." *Social Studies of Science* 37(6): 827–853.

Evans, T., and J. Harris. 2004. "Street-Level Bureaucracy, Social Work, and the (Exaggerated) Death of Discretion." *British Journal of Social Work* 34(6): 871–895.

Fargion, S. 2006. "Thinking Professional Social Work: Expertise and Professional Ideologies in Social Workers' Accounts of Their Practice." *Journal of Social Work* 6(3): 255–273.

Fischer, J. 1993. "Empirically Based Practice: The End of Ideology?" In *Single-System Designs in the Social Services: Issues and Options for the 1990s*, ed. M. Bloom. Binghamton, N.Y.: Haworth.

Flick, U. 2006. *An Introduction to Qualitative Research*. London: Sage.

Forster, E. M. 1954. *Collected Short Stories*. Harmondsworth: Penguin.

Fortune, A. E. 2012. "Development of the Task-Centered Model." In *From Task-Centered Social Work to Evidence-Based and Integrative Practice: Reflections on History and Implementation*, ed. T. L. Rzepnicki, S. G. McCracken, and H. E. Briggs. Chicago: Lyceum.

Fortune, A. E. 2014. "How Quickly We Forget: Comments on 'A Historical Analysis of Evidence-Based Practice in Social Work: The Unfinished Journey Towards an Empirically Grounded Profession.'" *Social Service Review* 88(2): 217–223.

Fortune, A., P. McCallion, and K. Briar-Lawson. 2010. "Building Evidence-Based Intervention Models." In *Social Work Practice Research for the Twenty-First Century*, ed. A. Fortune, P. McCallion, and K. Briar-Lawson, 279–295. New York: Columbia University Press.

Foucault, M. 1973. *Birth of the Clinic*. London: Routledge.

Foucault, M. 1991. "What Is an Author?" In *The Foucault Reader*, ed. P. Rabinow. London: Penguin.

Foucault, M. 2002. *The Archaeology of Knowledge*. London: Routledge.

Fox, R. 2012. "Professional Voices from the Field." In *Social worker as Researcher: Integrating Research with Advocacy*, ed. T. Maschi and R. Youdin. Boston: Pearson Online, in Pearson's "My Social Work Lab." http://www.pearsonhighered.com/product?ISBN=0205594948#tabbed.

Froggett, L., and S. Briggs. 2012. "Practice-Near and Practice-Distant Methods in Human Services Research." *Journal of Research Practice* 8(2), article M9. http://jrp.icaap.org/index.php/jrp/article/view/318/276.

Fuller, S. 2009. *The Sociology of Intellectual Life*. London: Sage.

Gal, J., and M. Ajzenstadt. 2013. "The Long Path from Soup Kitchens to a Welfare State in Israel." *Journal of Policy History* 25(2): 240–263.

Gambrill, E. 2010. "Evidence-Informed Practice: Antidote to Propaganda in the Helping Professions?" *Research on Social Work Practice* 20(3): 302–320.

Garrett, P. M. 2013. *Social Work and Social Theory: Making Connections*. Bristol: Policy.

Garrison, J. 1988. "The Impossibility of Atheoretical Research." *Journal of Educational Thought* 22(1): 21–25.

Geertz, C. 1973. "Thick Description: Towards an Interpretive Theory of Culture." In *The Interpretation of Cultures*. New York: Basic Books.

Geertz, C. 1983. "Common Sense as a Cultural System." In *Local Knowledge*. New York: Basic Books.

Gelsthorpe, L. 1992. "Response to Martin Hammersley's Paper on Feminist Methodology." *Sociology* 26(2): 213–218.

Gergen, K. 2009. *An Invitation to Social Construction*. Thousand Oaks, Calif.: Sage.

Gibbs, L., and E. Gambrill. 2002. "Evidence-Based Practice—Counterarguments to Objections." *Research on Social Work Practice* 12(3): 452–476.

Giddens, A. 1993. *New Rules of Sociological Method*. Stanford, Calif.: Stanford University Press.

Gilgun, J. 2005. "Qualitative Research and Family Psychology." *Journal of Family Psychology* 19 (1): 40–50.

Godin, B., and Y. Gingras. 2002. "The Experimenters' Regress: From Scepticism to Argumentation." *Studies in History and Philosophy of Science* 33: 137–152.

Grau, C. 2007. "Bad Dreams, Evil Dreams, and the Experience Machine: Philosophy and *The Matrix*." In *Introduction to Philosophy: Classical and Contemporary Readings*, ed. J. Perry et al., 195–202. New York: Oxford University Press.

Graybeal, C. T. 2007. "Evidence for the Art of Social Work." *Families in Society* 88(4): 513–523.

Greene, J. 1996. "Qualitative Evaluation and Scientific Citizenship." *Evaluation* 2(3): 277–289.

Greenspan, N. T. 2005. *The End of the Certain World*. Chichester: John Wiley and Sons.

Greenwood, E. 1945. *Experimental Sociology: A Study in Method*. New York: King's Crown.

Gregory, R. L. 1987. *The Oxford Companion to the Mind*. Oxford: Oxford University Press.

Grisso, T., and G. M. Vincent. 2005. "The Empirical Limits of Forensic Mental Health Assessment." *Law and Human Behavior* 29(1): 1–5.

Grossberg, L. 1996. "On Postmodernism and Articulation: An Interview with Stuart Hall." In *Stuart Hall: Critical Dialogues in Cultural Studies*, ed. D. Morley and K. H. Chen. London: Routledge.

Guba, E. G. 1990. "The Alternative Paradigm Dialog." In *The Paradigm Dialog*, ed. E. G. Guba. Newbury Park, Calif.: Sage.

Hackett, E. J. 2005. "Introduction. Special Guest Issue on Scientific Collaboration." *Social Studies of Science* 35(5): 667–671.

Hacking, I. 1991. "The Making and Molding of Child Abuse." *Critical Inquiry* 17(2): 253–288.

Hakim, C. 1987. *Research Design*. London: Unwin Hyman.

Hall, T. 2001. "Caught Not Taught: Ethnographic Research at a Young People's Accommodation Project." In *Qualitative Research in Social Work*, ed. I. Shaw and N. Gould. London: Sage.

Hammersley, M. 1992. *What's Wrong with Ethnography?* London: Routledge.

Hammersley, M. 2005. "Is the Evidence-Based Practice Movement Doing More Harm Than Good? Reflections on Iain Chalmers's Case for Research-Based Policy Making and Practice." *Evidence and Policy* 1(1): 85–100.

Hammersley, M. 2008. "The Critical Case of Analytic Induction." In *Questioning Qualitative Inquiry: Critical Essays*. London: Sage.

Hammersley, M. 2012. *What Is Analytic Induction?* In NCRM Research Methods Festival 2012, July 2–5, 2012, St. Catherine's College, Oxford. http://eprints.ncrm.ac.uk/2763/.

Hammersley, M., and R. Gomm. 1997. "Bias in Social Research." *Sociological Research Online* 2(1). http://www.socresonline.org.uk/socresonline/2/1/2.html.

Hanson, N. R. 1958. *Patterns of Discovery: An Inquiry Into the Conceptual Foundations of Science*. Cambridge: Cambridge University Press.

Harding, S., ed. 1987. *Feminism and Methodology*. Buckingham: Open University Press.

Harré, R. 1989. "Language Games and Texts of Identity." In *Texts of Identity*, ed. J. Shotter and K. J. Gergen. London: Sage.

Harrison, P. 2007. *The Fall of Man and the Foundations of Science*. Cambridge: Cambridge University Press.

Hartsock, N. 1983. "The Feminist Standpoint: Developing the Ground for a Specifically Feminist Historical Materialism." In *Discovering Reality*, ed. S. Harding and M. Hintikka. Dordrecht: Reidl.

Hartswood, M., R. Proctor, R. Slack, et al. 2002. "Corealisation: Towards a Principled Synthesis of Ethnomethodology and Participatory Design." *Scandinavian Journal of Information Systems* 14(2): 9–30.

Harvey, L. 1990. *Critical Social Research*. London: Unwin Hyman.

Hawkesworth, M. 1989. "Knowers, Knowing, Known: Feminist Theory and Claims of Truth." *Signs: Journal of Women in Culture and Society* 14(3): 533–555.

Healy, D., M. Harris, D. Cattell, et al. 2005. "Service Utilization in 1896 and 1996: Morbidity and Mortality Data from North Wales." *History of Psychiatry* 16(1): 27–41.

Heidegger, M. 1954/1977. "The Question Concerning Technology." In *The Question Concerning Technology and Other Essays*. New York: Harper and Row.

Hekman, S. 1990. *Gender and Knowledge: Elements of a Postmodern Feminism*. Cambridge: Policy.

Helm, P., ed. 1987. *Objective Knowledge: A Christian Perspective.* Leicester: Inter-Varsity.
Helm, P. 2014. *Faith, Form, and Fashion: Classical Reformed Theology and Its Modern Critics.* Eugene, Ore.: Cascade.
Hingley-Jones, H. 2009. "Developing Practice-Near Social Work Research to Explore the Emotional Words of Severely Learning Disabled Adolescents in 'Transition' and Their Families." *Journal of Social Work Practice* 23(4): 413–428.
Hitchings, H. 2008. *The Secret Life of Words: How English Became English.* London: John Murray.
Hooykaas, R. 1957. *Christian Faith and the Freedom of Science.* London: Tyndale.
Hooykaas, R. 1960. *The Christian Approach in Teaching Science.* London: Tyndale.
Hooykaas, R. 1972. *Religion and the Rise of Modern Science.* Edinburgh: Scottish Academic Press.
Horkheimer, M. 2002. *Critical Theory: Selected Essays.* New York: Continuum.
House, E. 1980. *Evaluating with Validity.* Beverly Hills, Calif.: Sage.
House, E. 1991. "Evaluation and Social Justice: Where Are We Now?" In *Evaluation and Education: At Quarter Century*, ed. M. McLaughlin and D. Phillips. Chicago: University of Chicago Press.
House, E. 1993. *Professional Evaluation.* Newbury Park, Calif.: Sage.
House, E. 1995. "Putting Things Together Coherently: Logic and Justice." In *Reasoning in Evaluation: Inferential Links and Leaps*, ed. D. Fournier. San Francisco: American Evaluation Association/Jossey-Bass.
House, E., and K. Howe. 1999. *Evaluation and Values.* Thousand Oaks, Calif.: Sage.
Howard, M. O., C. J. McMillen, and D. E. Pollio. 2003. "Teaching Evidence-Based Practice: Towards a New Paradigm for Social Work Education." *Research on Social Work Practice* 13(2): 234–259.
Hudson, B. 2004. "Analyzing Network Partnerships." *Public Management Review* 6(1): 75–94.
Hughes, E. C. 1984. *The Sociological Eye.* New Brunswick, N.J.: Transaction.
Hughes, J. 1980. *The Philosophy of Social Research.* Harlow: Longman Group.
Imre, R. W. 1985. "Tacit Knowledge in Social Work Research and Practice." *Smith College Studies in Social Work* 55(2): 137–149.
Irvine, E. 1969. "Education for Social Work: Science or Humanity?" *Social Work* 26(4): 3–6.
Jacob, E. 1987. "Qualitative Research Traditions: A Review." *Review of Educational Research* 57(1): 1–50.
Jacob, M. C. 1992. "Science and Politics in the Late Twentieth Century." *Social Research* 59(3): 487–503.
Jones, H. 1962. *Crime and the Penal System.* London: University Tutorial Press.

Kahlberg, S. 2000. "Max Weber." In *The Blackwell Companion to Major Social Theorists,* ed. G. Ritzer, 144–204. Malden, Mass.: Blackwell.

Kemp, S. 2012. "Evaluating Interests in Social Science: Beyond Objectivist Evaluation and the Nonjudgmental Stance." *Sociology* 46(4): 664–679.

Kirk, S., and W. J. Reid. 2002. *Science and Social Work.* New York: Columbia University Press.

Kuhn, T. S. 1970. *The Structure of Scientific Revolutions.* Chicago: University of Chicago Press.

Kuhn, T. S. 1992. "The Trouble with the Historical Philosophy of Science." Robert and Maurine Rothschild Distinguished Lecture, Department of the History of Science, Harvard University.

Kushner, S. 1996. "The Limits of Constructivism in Evaluation." *Evaluation* 2(2): 189–200.

Lakatos, I. 1970. "Falsification and the Methodology of Scientific Research Programmes." In *Criticism and the Growth of Knowledge,* ed. I. Lakatos and A. Musgrave. Cambridge: Cambridge University Press.

Lather, P. 1986a. "Issues of Validity in Openly Ideological Research." *Interchange* 17(4): 63–84.

Lather, P. 1986b. "Research as Praxis." *Harvard Educational Review* 56(3): 257–277.

Lather, P. 1991. *Getting Smart: Feminist Research and Pedagogy with/in the Postmodern.* New York: Routledge.

Latour, B. 1992. "One More Turn After the Social Turn: Easing Science Studies Into the Nonmodern World." In *The Social Dimensions of Science,* ed. E. McMullin, 272–292. Notre Dame: Notre Dame University Press.

Laurent, V. 2008. "ICT and Social Work: A Question of Identities?" In *The Future of Identity in the Information Society,* ed. S. Fischer-Hubner et al., 375–386. Boston: Springer.

Lazega, E. 1997. "Network Analysis and Qualitative Research: A Method of Contextualization." In *Context and Method in Qualitative Research,* ed. G. Miller and R. Dingwall. London: Sage.

Lee, R. 2004. "Recording Technologies and the Interview in Sociology, 1920–2000." *Sociology* 38(5): 869–889.

Lengermann, P., and G. Niebrugge. 2007. "Thrice Told: Narratives of Sociology's Relation to Social Work." In *Sociology in America: A History,* ed. C. Calhoun, 63–114. Chicago: University of Chicago Press.

Leplin, J., ed. 1984. *Scientific Realism.* Berkeley: University of California Press.

Lorenz, W. 2004. "Research as an Element in Social Work's Ongoing Search for Identity." In *Reflecting on Social Work—Discipline and Profession,* ed. J. Powell, R. Lovelock, and K. Lyons. Aldershot: Ashgate.

Lorenz, W. 2007. "Practising History: Memory and Contemporary Professional Practice." *International Social Work* 50(5): 597–612.

Lubove, R. 1965. *The Professional Altruist.* Cambridge, Mass.: Harvard University Press.

Lyman, S. M., and M. B. Scott. 1970. *A Sociology of the Absurd.* New York: Appleton Century-Crofts.

Macdonald, G. 1990. "Allocating Blame in Social Work." *British Journal of Social Work* 20(6): 525–546.

Macdonald, G. 2000a. "Critical Thinking." In *The Blackwell Encyclopedia of Social Work,* ed. M. Davies. Oxford: Blackwell.

Macdonald, G. 2000b. "Evidence-Based Practice." In *The Blackwell Encyclopedia of Social Work,* ed. M. Davies. Oxford: Blackwell.

Macdonald, M. 1966. "Reunion at Vocational High." *Social Service Review* 40(2): 175–189.

MacIver, R. M. 1931. *The Contribution of Sociology to Social Work.* New York: Columbia University Press.

MacKay, D. 1981. "Value-Free Knowledge: Myth or Norm?" *Christian Graduate* 34(2): 13–17.

MacKay, D. 1987. "Objectivity as a Christian Value." In *Objective Knowledge: A Christian Perspective,* ed. P. Helm. Leicester: Inter-varsity.

MacKenzie, D. A. 1981. *Statistics in Britain, 1865–1930: The Social Construction of Scientific Knowledge.* Edinburgh: Edinburgh University Press.

Madge, J. 1953. *The Tools of Social Science.* London: Longmans Green.

Madge, J. 1963. *The Origins of Scientific Sociology.* London: Tavistock.

Martin, M. C. 2007. "Crossing the Line: Observations from East Detroit, Michigan, USA." *Qualitative Social Work* 7(4): 465–475.

Martinez-Brawley, E. E., and P. M.-B. Zorita. 2007. "Tacit and Codified Knowledge in Social Work: A Critique of Standardization in Education and Practice." *Families in Society* 88(4): 534–542.

Massialas, B. G., and S. A. Jarra. 1983. *Education in the Arab World.* New York: Praeger.

Matthewman, S., and D. Hoey. 2006. "What Happened to Postmodernism?" *Sociology* 40(3): 529–547.

Mayer, J., and N. Timms. 1970. *The Client Speaks.* London: Routledge and Kegan Paul.

McClean, S., and A. Shaw. 2005. "From Schism to Continuum? The Problematic Relationship Between Expert and Lay Knowledge." *Qualitative Health Research* 15(6):729–749.

McCrea, K. T. 2007. "Makeovers for 'Bewhiskered' Research: Diverse Heuristics for Understanding Causality in the Social and Behavioural Sciences." In *Social Work Dialogues: Transforming the Canon in Inquiry, Practice, and Education,* ed. S. L. Witkin and D. Saleebey. Alexandria, Va.: Council on Social Work Education.

McLennan, G. 1995. "Feminism, Epistemology, and Postmodernism: Reflections on Current Ambivalence." *Sociology* 29(2): 391–409.

Mead, G. H. 1923. "Scientific Method and the Moral Sciences." *International Journal of Ethics* 33(3): 229–247.

Medawar, P. 1984. *The Limits of Science*. Oxford: Oxford University Press.

Merton, R. K. 1971. "Social Problems and Sociological Theory." In *Contemporary Social Problems*, ed. R. Merton and R. Nisbet. New York: Harcourt Brace Jovanovich.

Meyer, H. J., E. F. Borgatta, and W. C. Jones. 1965. *Girls at Vocational High*. New York: Russell Sage Foundation.

Milford Conference. 1929. *Social Case Work: Generic and Specific*. Washington, D.C.: American Association of Social Workers.

Mills, C. W. 1943. "The Professional Ideology of Social Pathologists." *American Journal of Sociology* 49(2): 165–180.

Mills, C. W. 1963. "On Knowledge and Power." In *Power, Politics, and People*. New York: Oxford University Press.

Mills, C. W. 2008. "Sociological Poetry." In *The Politics of Truth: Selected Writings of C. Wright Mills*, ed. J. Summers. Oxford: Oxford University Press.

Mullen, E. J. 2004. "Facilitating Practitioner Use of EBP." In *Evidence-Based Practice Manual: Research and Outcome Measures in Health and Human Services*, ed. A. R. Roberts and K. R. Yeager. New York: Oxford University Press.

Mullen, E. J. 2015. "Reconsidering the 'Idea' of Evidence in Evidence-Based Policy and Practice." DOI: 10.1080/13691457.2015.1022716.

Mullen, E. J., and W. Bacon. 2004. "Implementation of Practice Guidelines and Evidence-Based Treatment." In *Evidence-Based Practice Manual: Research and Outcome Measures in Health and Human Services*, ed. A. R. Roberts and K. R. Yeager. New York: Oxford University Press.

Munson, C. E. 2004. "Evidence-Based Treatment for Traumatized and Abused Children." In *Evidence-Based Practice Manual: Research and Outcome Measures in Health and Human Services*, ed. A. R. Roberts and K. R. Yeager. New York: Oxford University Press.

Neander, K., and C. Skott. 2006. "Important Meetings with Important Persons: Narratives from Families Facing Adversity and Their Key Figures." *Qualitative Social Work* 5(3): 295–311.

Neander, K., and C. Skott. 2008. "Bridging the Gap—The Co-creation of a Therapeutic Process: Reflections by Parents and Professionals on Their Shared Experiences of Early Childhood Interventions." *Qualitative Social Work* 7(3): 289–309.

Nicholson, L., ed. 1990. *Feminism/Postmodernism*. London: Routledge.

Nokes, P. 1967. *The Professional Task in Welfare Practice*. London: Routledge and Kegan Paul.

Nowotny, H. 2003. "Dilemma of Expertise: Democratising Expertise and Socially Robust Knowledge." *Science and Public Policy* 30(3): 151–156.

Nurius, P., and S. Kemp. 2012. "Social Work, Science, and Social Impact: Crafting an Integrative Conversation." *Research on Social Work Practice* 22(5): 548–552.

Oakley, A. 1998a. "Experimentation and Social Interventions: A Forgotten but Important History." *British Medical Journal* 317: 1239–1242.

Oakley, A. 1998b. "Public Policy Experimentation: Lessons from America." *Policy Studies* 19(2): 93–114.

Oakley, A. 1999. "Paradigm Wars: Some Thoughts on a Personal and Public Trajectory." *International Journal of Social Research Methodology* 2(3): 247–254.

Oakley, A. 2000. *Experiments in Knowing: Gender and Method in the Social Sciences.* Oxford: Polity.

Oakley, A. 2011. *A Critical Woman: Barbara Wootton, Social Science, and Public Policy in the Twentieth Century*. London: Bloomsbury.

Oakley, A. 2014. *Father and Daughter: Patriarchy, Gender, and Social Science*. Bristol: Policy.

Oancea, A., and J. Furlong. 2007. "Expressions of Excellence and the Assessment of Applied and Practice-Based Research." *Research Papers in Education* 22(2): 119–137.

O'Farrell, C. 2005. *Michel Foucault*. London: Sage.

Okamato, S. K., and C. W. LeCroy. 2004. "Evidence-Based Practice and Manualized Treatment with Children." In *Evidence-Based Practice Manual: Research and Outcome Measures in Health and Human Services*, ed. A. R. Roberts and K. R. Yeager. New York: Oxford University Press.

Okpych, N. J., and J. L.-H. Yu. 2014. "A Historical Analysis of Evidence-Based Practice in Social Work: The Unfinished Journey Toward an Empirically Grounded Profession." *Social Service Review* 88(1): 3–58.

Olesen, V. 1994. "Feminisms and Models of Qualitative Research." In *Handbook of Qualitative Methodology*, ed. N. Denzin and Y. Lincoln. Thousand Oaks, Calif.: Sage.

Orme, J. 2003. "It's Feminist Because I Say So! Feminism, Social Work, and Critical Practice in the UK." *Qualitative Social Work* 2(2): 131–154.

Packer, M. 2011. *The Science of Qualitative Research*. New York: Cambridge University Press.

Padgett, D. 2008. *Qualitative Methods in Social Work Research*. Thousand Oaks, Calif.: Sage.

Palinkas, L. 2014. "Causality and Causal Inference in Social Work: Quantitative and Qualitative Perspectives." *Research on Social Work Practice* 24(5): 540–547.

Park, R. E. 1929. "The City as a Social Laboratory." In *Chicago: An Experiment in Social Science Research*, ed. T. V. Smith and L. D. White. Chicago: University of Chicago Press.

Parton, N. 2008a. "Towards the Preventative-Surveillance State: The Current Changes in Children's Services in England." In *Child Protection and Welfare Social Work: Contemporary Themes and Practice Perspectives*, ed. K. Burns and D. Lynch. Dublin: A. & A. Farmar.

Parton, N. 2008b. "Changes in the Form of Knowledge in Social Work: From the 'Social' to the 'Informational'?" *British Journal of Social Work* 38(2): 253–269.

Parton, N., and S. Kirk. 2010. "The Nature and Purposes of Social Work." In *Sage Handbook of Social Work*, ed. I. Shaw et al. London: Sage.

Patton, M. Q. 1988. "The Evaluator's Responsibility for Utilization." *Evaluation Practice* 9(1): 5–24.

Patton, M. Q. 1996. *Utilization Focussed Evaluation*. Thousand Oaks, Calif.: Sage.

Pawson, R., A. Boaz, L. Grayson, et al. 2003. *Types and Quality of Knowledge in Social Care*. Knowledge Review 3. London: Social Care Institute for Excellence.

Pawson, R., G. Wong, and L. Owen. 2011. "Known Knowns, Known Unknowns, Unknown Unknowns: The Predicament of Evidence-Based Policy." *American Journal of Evaluation* 32(4): 518–546.

Phillips, C. 2007. "Pain(ful) Subjects: Regulated Bodies in Medicine and Social Work." *Qualitative Social Work* 6(2): 197–212.

Phillips, C., and I. Shaw. 2011. "Innovation in Social Work Research." *British Journal of Social Work* 41(4): 609–624.

Phillips, D. 1990a. "Postpositivistic Science: Myths and Realities." In *The Paradigm Dialog*, ed. E. Guba. Newbury Park, Calif.: Sage.

Phillips, D. 1990b. "Subjectivity and Objectivity: An Objective Inquiry." In *Qualitative Inquiry in Education*, ed. E. Eisner and A. Peshkin. New York: Teachers College, Columbia University.

Phillips, D. C. 2000. *The Expanded Social Scientist's Bestiary*. Lanham, Md.: Rowan and Littlefield.

Philp, M. 1979. "Notes on the Form of Knowledge in Social Work." *Sociological Review* 27(1): 83–111.

Pickering, M., and E. Keightley. 2006. "The Modalities of Nostalgia." *Current Sociology* 54(6): 919–941.

Pithouse, A., and P. Atkinson. 1988. "Telling the Case: Occupational Narrative in a Social Work Office." In *Styles of Discourse*, ed. N. Coupland. London: Croom Helm.

Pithouse, A., C. Hall, S. Peckover, and S. White. 2009. "A Tale of Two CAFs: The Impact of the Electronic Common Assessment Framework." *British Journal of Social Work* 39(4): 599–612.

Platt, J. 1996. *A History of Sociological Research Methods in America, 1920–1960*. Cambridge: Cambridge University Press.

Platt, J. R. 1964. "Strong Inference." *Science* n.s. 146(3642): 347–353.

Plummer, K. 1997. "Introducing Chicago Sociology." In *The Chicago School: Critical Assessments*, vol. 1: *A Chicago Canon?*, ed. K. Plummer. London: Routledge.
Polanyi, M. 1966. *The Tacit Dimension*. Chicago: University of Chicago Press.
Polkinghorne, D. E. 2004. *Practice and the Human Sciences: The Case for a Judgment-Based Practice of Care*. Albany, N.Y.: SUNY Press.
Popay, J., C. Thomas, G. Williams, et al. 2003. "A Proper Place to Live: Health Inequalities, Agency, and the Normative Dimensions of Space." *Social Science and Medicine* 57(1): 55–65.
Popkewitz, T. 1990. "Whose Future? Whose Past?" In *The Paradigm Dialog*, ed. E. Guba. Newbury Park, Calif.: Sage.
Popper, K. 1966. *The Open Society and Its Enemies*. Vol. 2. London: Routledge and Kegan Paul.
Popper, K. 2002. *The Logic of Scientific Discovery*. London: Routledge.
Prior, L. 2003. "Belief, Knowledge, and Expertise: The Emergence of the Lay Expert in Medical Sociology: Belief, Knowledge, and Expertise." *Sociology of Health and Illness* 25(3): 41–57.
Proctor, E. K., and A. Rosen. 2004. "Concise Standards for Developing Evidence-Based Practice Guidelines." In *Evidence-Based Practice Manual: Research and Outcome Measures in Health and Human Services*, ed. A. R. Roberts and K. R. Yeager. New York: Oxford University Press.
Queen, S. A. 1928. "Social Interaction in the Interview: An Experiment." *Social Forces* 6(4): 545–558.
Ragin, C. C. 1989. *The Comparative Method: Moving Beyond Qualitative and Quantitative Strategies*. Oakland: University of California Press.
Ragin, C. C. 1997. "Turning the Tables: How Case-Oriented Research Challenges Variable-Oriented Research." *Comparative Social Research* 16(1): 27–42.
Raynor, P. 2003. "Evidence-Based Probation and Its Critics." *Probation Journal* 50(4): 334–345.
Reid, W. J. 1988. "Service Effectiveness and the Social Agency." In *Managing for Effectiveness in Social Welfare Organizations*, ed. R. Patti et al. New York: Haworth.
Reid, W. J. 1994a. "The Empirical Practice Movement." *Social Service Review* 68(2): 165–184.
Reid, W. J. 1994b. "Reframing the Epistemological Debate." In *Qualitative Research in Social Work*, ed. E. Sherman and W. J. Reid. New York: Columbia University Press.
Reid, W. J. 2001. "The Role of Science in Social Work: The Perennial Debate." *Journal of Social Work* 1(3): 273–293.
Reid, W. J., and A. W. Shyne. 1969. *Brief and Extended Casework*. New York: Columbia University Press.
Reid, W. J., and P. Zettergren. 1999. "A Perspective on Empirical Practice." In *Evaluation and Social Work Practice*, ed. I. Shaw and J. Lishman. London: Sage.

Reisch, M., and C. Garvin. 2015. *Social Work Practice and Social Justice: Concepts, Challenges, and Strategies*. New York: Oxford University Press.

Richards, R. J. 1981. "Natural Selection and Other Models in the Historiography of Science." In *Scientific Inquiry and the Social Sciences*, ed. M. Brewer and B. Collins. San Francisco: Jossey Bass.

Riesman, D., and H. Becker. 1984. "Introduction." In *The Sociological Eye: Selected Papers*, by E. C. Hughes. New Brunswick, N.J.: Transaction.

Riessman, C. K. 2001. "Personal Troubles as Social Issues: A Narrative of Infertility in Context." In *Qualitative Research in Social Work*, ed. I. Shaw and N. Gould. London: Sage.

Rip, A. 2002. "Science for the Twenty-First Century." In *The Future of the Sciences and Humanities*, ed. P. Tindemans et al. Amsterdam: Amsterdam University Press.

Roberts, A. R., and K. R. Yeager, eds. 2004. *Evidence-Based Practice Manual: Research and Outcome Measures in Health and Human Services*. New York: Oxford University Press.

Robinson, W. S. 1951. "The Logical Structure of Analytic Induction." *American Sociological Review* 16(6): 812–818.

Rodriguez, N., and A. Ryave. 2002. *Systematic Self Observation*. Thousand Oaks, Calif.: Sage.

Rogers, C. 1965. *Client-Centered Therapy*. Boston: Houghton Mifflin.

Romm, N. 1995. "Knowing as an Intervention." *Systems Practice* 8(2): 137–167.

Romm, N. 1996. "Inquiry-and-Intervention in Systems Planning: Probing Methodological Rationalities." *World Futures* 47: 25–36.

Rooney, R. H. 2010. "Task-Centered Practice in the United States." In *Social Work Practice Research for the Twenty-First Century*, ed. A. Fortune et al., 195–202. New York: Columbia University Press.

Ross, P. 2010. "Problematizing the User in User-Centered Production: A New Media Lab Meets Its Audiences." *Social Studies of Science* 41(2): 251–270.

Ruckdeschel, R., and A. Chambon. 2010. "The Uses of Social Work Research." In *The Sage Handbook of Social Work Research*, ed. I. Shaw et al. London: Sage.

Russell, B. 1913. "Science as an Element in Culture." *New Statesman* (May 24, 1913).

Sackett, D., W. Rosenberg, J. Gray, et al. 1996. "Evidence-Based Medicine: What It Is and What It Isn't." *British Medical Journal* 312(7023): 71–72.

Scheff, T. 1997. "Part/Whole Morphology: Unifying Single Case and Comparative Methods." *Sociological Research Online* 2.3.1.

Schivelbusch, W. 1977. *The Railway Journey: The Industrialization of Time and Space in the Nineteenth Century*. Los Angeles: University of California Press.

Schmidt, M. 1993. "Grout: Alternative Kinds of Knowledge and Why They Are Ignored." *Public Administration Review* 53(6): 525–530.

Schön, D. 1983. *The Reflective Practitioner: How Professionals Think in Action.* New York: Basic Books.
Schön, D. 1992. "The Crisis of Professional Knowledge and the Pursuit of an Epistemology of Practice." *Journal of Interprofessional Care* 6(1): 49–63.
Schuerman, J. R. 1987. "Passion, Analysis, and Technology." *Social Service Review* 61(1): 3–18.
Schutz, A. 1967a. "Common-Sense and Scientific Interpretation of Human Action." In *Collected Papers*, vol. 1: *The Problem of Social Reality.* The Hague: Martinus Nijhoff.
Schutz, A. 1967b. "The Well-Informed Citizen: A Study in the Social Distribution of Knowledge." In *Collected Papers*, vol. 2: *Studies in Social Theory.* The Hague: Martinus Nijhoff.
Schutz, A. 1967c. "The Stranger: An Essay in Social Psychology." In *Collected Papers*, vol. 2: *Studies in Social Theory.* The Hague: Martinus Nijhoff.
Schutz, A. 1967d. "On Multiple Realities." In *Collected Papers*, vol. 1: *The Problem of Social Reality.* The Hague: Martinus Nijhoff.
Schutz, A. 1967e. "Concept and Theory Formation." In *Collected Papers*, vol. 1: *The Problem of Social Reality.* The Hague: Martinus Nijhoff.
Schutz, A. 1967f. "Construct of Thought Objects by the Social Sciences." In *Collected Papers*, vol. 1: *The Problem of Social Reality.* The Hague: Martinus Nijhoff.
Schwab, J. 1969. "The Practical: A Language for Curriculum." *School Review* (November): 1–23.
Schwandt, T. 1993. "Theory for the Moral Sciences: Crisis of Identity and Purpose." In *Theory and Concepts in Qualitative Research: Perspectives from the Field*, ed. D. Flinders and G. Mills. New York: Teachers College Press.
Schwandt, T. 1999. "On Understanding Understanding." *Qualitative Inquiry* 5(4): 451–464.
Scriven, M. 1986. "New Frontiers of Evaluation." *Evaluation Practice* 7(1): 7–44.
Scriven, M. 1997. "Truth and Objectivity in Evaluation." In *Evaluation for the Twenty-First Century*, ed. E. Chelimsky and W. Shadish. Thousand Oaks, Calif.: Sage.
Seale, C. 1999. *The Quality of Qualitative Research.* London: Sage.
Seed, P. 1973. *The Expansion of Social Work in Britain.* London: Routledge and Kegan Paul.
Shadish, W., T. Cook, and L. Leviton. 1990. *Foundations of Program Evaluation: Theories of Practice.* Newbury Park, Calif.: Sage.
Shapin, S. 1994. *A Social History of Truth.* Chicago: University of Chicago Press.
Shapin, S. 2012. "The Sciences of Subjectivity." *Social Studies of Science* 42(2): 170–184.
Shaw, I. 1996. *Evaluating in Practice.* Aldershot: Ashgate.

Shaw, I. 1999. "Seeing the Trees for the Wood: The Politics of Practice Evaluation." In *The Politics of Research and Evaluation in Social Work*, ed. B. Broad. Birmingham: Venture.

Shaw, I. 2007. "Is Social Work Research Distinctive?" *Social Work Education* 26(7): 659–669.

Shaw, I. 2009. "Rereading *The Jack-Roller*: Hidden Histories in Sociology and Social Work." *Qualitative Inquiry* 15(7): 1241–1264.

Shaw, I. 2010. "Places in Time: Contextualizing Social Work Research." In *Sage Handbook of Social Work Research*, ed. I. Shaw et al. London: Sage.

Shaw, I. 2011a. *Evaluating in Practice*. 2nd ed. Aldershot: Ashgate.

Shaw, I. 2011b. "Social Work Research: An Urban Desert?" *European Journal of Social Work* 14(1): 11–26.

Shaw, I. 2012a. *Practice and Research*. Aldershot: Ashgate.

Shaw, I. 2012b. "The Positive Contributions of Quantitative Methodology to Social Work Research: A View from the Sidelines." *Research on Social Work Practice* 22(2): 129–134.

Shaw, I. 2012c. "Ways of Knowing in Social Work." In *Social Work Theory and Methods*, 2nd ed., ed. M. Gray and S. Webb. London: Sage.

Shaw, I. 2013. "Angels and Devils the Following Day." *Qualitative Social Work* 12(2): 234–244.

Shaw, I. 2014a. "Sociology and Social Work: In Praise of Limestone?" In *The Palgrave Handbook of Sociology in Britain*, ed. J. Holmwood and J. Scott. London: Palgrave.

Shaw, I. 2014b. "Noel Timms: A Brief Appreciation." *Qualitative Social Work* 13(6): 742–748.

Shaw, I. 2014c. "Before I Built a Wall: One Sort of Dialogue." *Qualitative Social Work* 13(2): 181–183.

Shaw, I. 2014d. "A Science of Social Work?—Response to John Brekke." *Research on Social Work Practice* 24(5): 524–526.

Shaw, I. 2015a. "The Archaeology of Research Practices: A Social Work Case." *Qualitative Inquiry* 21(1): 36–49.

Shaw, I. 2015b. "Sociological Social Workers: A History of the Present?" *Nordic Journal of Social Work Research* 5, sup 1: 7–24.

Shaw, I. Forthcoming. "Case work: Re-forming the Relationship Between Sociology and Social Work." *Qualitative Research*. DOI: 10.1177/1468794114567497.

Shaw, I., and K. Al-Awwad. 1994. "Culture and the Indigenization of Quality in Third World Social Research." *Journal of Social Development in Africa* 9(1): 59–71.

Shaw I., H. Arksey, and A. Mullender. 2006. "Recognizing Social Work." *British Journal of Social Work* 36(2): 227–246.

Shaw, I., and S. Holland. 2014. *Doing Qualitative Research in Social Work*. London: Sage.

Shaw, I., J. Marsh, and M. Hardy, eds. 2015. *Social Work Research*. London: Sage.

Shaw, I. and M. Norton. 2007. *Kinds and Quality of Social Work Research in Higher Education*. London: Social Care Institute for Excellence. http://www.scie.org.uk/publications/reports/report17.asp.

Shaw, I., and A. Shaw. 1997. "Game Plans, Buzzes, and Sheer Luck: Doing Well in Social Work." *Social Work Research* 21(2): 69–79.

Shaw, I., M. Bell, I. Sinclair, et al. 2009. "An Exemplary Scheme? An Evaluation of the Integrated Children's System." *British Journal of Social Work* 39(4): 613–626.

Shaw, I., K. Morris, and A. Edwards. 2009. "Technology, Social Services, and Organizational Innovation; or, How Great Expectations in London and Cardiff are Dashed in Lowestoft and Cymtyrch." *Journal of Social Work Practice* 23 (4): 383–400.

Sheffield, Ada E. 1920. *The Social Case History: Its Construction and Content*. New York: Russell Sage Foundation.

Sheffield, Ada E. 1922. *Case-Study Possibilities, a Forecast*. Boston: Research Bureau on Social Case Work.

Shimuzu, H, and T. Hirao. 2009. "Interorganizational Collaborative Research Networks in Semiconductor Lasers, 1975–1994." *Social Science Journal* 46(2): 233–251.

Shyne, A. W. 1965. "An Experimental Study of Casework Methods." *Social Casework* 46 (November).

Sinclair, I. 2012. "Interview." *Qualitative Social Work* 11(2): 206–215.

Skagestad, P. 1981. "Hypothetical Realism." In *Scientific Inquiry and the Social Sciences*, ed. M. Brewer and B. Collins. San Francisco: Jossey Bass.

Skehill, C. 2004. *History of the Present of Child Protection and Welfare Social Work in Ireland* New York: Edward Mellen.

Skehill, C. 2007. "Researching the History of Social Work: Exposition of a History of the Present Approach." *European Journal of Social Work* 10(4): 449–463.

Small, A. W. 1896. "Scholarship and Social Agitation." *American Journal of Sociology* 1(5): 564–582.

Solomon, B. 2007. "Taking 'Guilty Knowledge' Seriously: Theorizing, Everyday Inquiry, and Action as 'Social Caretaking.'" In *Social Work Dialogues: Transforming the Canon in Inquiry, Practice, and Education*, ed. S. L. Witkin and D. Saleebey. Alexandria, Va.: Council on Social Work Education.

Sowa, J. F. 2005. "Theories, Models, Reasoning, Language, and Truth." http://www.jfsowa.com/logic/theories.htm.

Soydan, H. 1999. *The History of Ideas in Social Work*. Birmingham: Venture.

Specht, H., and M. Courtney. 1994. *Fallen Angels: How Social Work Has Abandoned Its Mission*. New York: Free Press.

Springer, D., N. Abell, and W. Hudson. 2002a. “Creating and Validating Rapid Assessment Instruments for Practice and Research: Part 1.” *Research on Social Work Practice* 12(3): 408–439.

Springer, D., N. Abell, and W. Hudson. 2002b. “Creating and Validating Rapid Assessment Instruments for Practice and Research: Part 2.” *Research on Social Work Practice* 12(4): 768–795.

Staller, K. 2007. “Metalogue as Methodology.” *Qualitative Social Work* 6(2): 137–157.

Staller, K. 2010. “Technology and Inquiry: Past, Present, and Future.” *Qualitative Social Work* 9 (2): 285–287

Staller, K. 2014. “Difficult Conversations: Talking *with* Rather Than *at*.” *Qualitative Social Work* 13(2): 167–175.

Stokes, D. E. 1997. *Pasteur’s Quadrant: Basic Science and Technological Innovation.* Washington, D.C.: Brookings Institution.

Swigonski, M. 1993. “Feminist Standpoint Theory and Questions of Social Work Research.” *Affilia* 8(2): 171–183.

Taylor, C. 1999. “Two Theories of Modernity.” *Public Culture* 11(1): 153–174.

Thyer, B. A. 2001. “Theory Testing in Social Work Research.” *Journal of Social Work Education* 37(1): 9–25.

Thyer, B. A. 2012. “The Scientific Value of Qualitative Research for Social Work.” *Qualitative Social Work* 11(2): 115–125.

Timms, N. 1968. *The Language of Social Casework.* London: Routledge and Kegan Paul.

Timms, N. 1972. “ . . . *and* Renoir *and* Matisse *and* . . . ” Inaugural Lecture. University of Bradford.

Timms, N. 2014. *A Sociological Approach to Social Problems.* London: Routledge.

Timms, N., and R. Timms. 1977. *Perspectives in Social Work.* London: Routledge and Kegan Paul.

Todd, A. J. 1919. *The Scientific Spirit and Social Work.* New York: Macmillan.

Uggerhøj, L. 2011. “What Is Practice Research in Social Work? Definitions, Barriers, and Possibilities.” *Social Work and Society* 9(1): 45–59.

Ulrich, W. 2001a. “A Philosophical Staircase for Information Systems Definition, Design, and Development: A Discursive Approach to Reflective Practice in ISD (Part 1).” *Journal of Information Technology Theory and Application* 3(3): 55–84.

Ulrich, W. 2001b. “Critically Systemic Discourse: A Discursive Approach to Reflective Practice in ISD (Part 2).” *Journal of Information Technology Theory and Application* 3(3): 85–106.

van Benthem, J. 2002. “Science and Society in Flux.” In *The Future of the Sciences and Humanities,* ed. P. Tindemans et al. Amsterdam: Amsterdam University Press.

van de Ven, A. H., and P. E. Johnson. 2006. “Knowledge for Theory and Practice.” *Academy of Management Review* 31(4): 802–821.

van Maanen, M. 2007. "Phenomenology of Practice." *Phenomenology and Practice* 1(1): 11–30.

van Maanen, J. 2011. *Tales of the Field: On Writing Ethnography*. Chicago: University of Chicago Press.

Videka, L., and J. A. Blackburn. 2010. "The Intellectual Legacy of William J Reid." In *Social Work Practice Research for the Twenty-First Century*, ed. A. Fortune et al., 183–194. New York: Columbia University Press.

Wallace, W. L. 2004. "The Logic of Science in Sociology." In *Social Research Methods: A Reader*, ed. C. Seale. London: Routledge.

Walter, I., S. Nutley, J. Percy-Smith, et al. 2004. *Improving the Use of Research in Social Care Practice*. London: SCIE. http://www.scie.org.uk/publications/knowledgereviews/kr07.asp.

Webb, B. 1929. *My Apprenticeship*. London: Longmans, Green and Co.

Webb, B. 1971. *My Apprenticeship*. Harmondsworth: Penguin.

Webber, M. 2014. "From Ethnography to Randomised Controlled Trial: An Innovative Approach to Developing Complex Social Interventions." *Journal of Evidence-Based Social Work* 11(1/2): 173–182.

Weber, M. 1948. "Science as a Vocation [1919]." In *From Max Weber: Essays in Sociology*, ed. H. H. Gerth and C. W. Mills, 129–156. London: Routledge and Kegan Paul.

Weiss, C. 1980. "Knowledge Creep and Decision Accretion." *Knowledge, Creation, Diffusion, Utilization* 1(3): 381–404.

Weiss, C. 1988. "If Programme Decisions Hinged Only on Information." *Evaluation Practice* 9(1): 15–28.

White, S. 2001. "Autoethnography as Reflexive Inquiry: The Research Act as Self-Surveillance." In *Qualitative Research in Practice*, ed. I. Shaw and N. Gould. London: Sage.

White, S., K. Broadhurst, D. Wastell, et al. 2009. "Whither Practice-Near Research in the Modernization Programme? Policy Blunders in Children's Services." *Journal of Social Work Practice* 23(4): 401–411.

Whitmore, E. 2001. "'People Listened to What We Had to Say': Reflections on an Emancipatory Qualitative Evaluation." In *Qualitative Research in Social Work*, ed. I. Shaw and N. Gould. London: Sage.

Williams, G. 1984. "The Genesis of Chronic Illness: Narrative Reconstruction." *Sociology of Health and Illness* 6(2): 175–200.

Williams, G., and J. Popay. 1994. "Lay Knowledge and the Privilege of Experience." In *Challenging Medicine*, ed. J. Gabe et al., 118–139. London: Routledge.

Williams, R. 1983. *Keywords: A Vocabulary of Culture and Society*. London: Fontana.

Witkin, S. 1999. "Constructing Our Future." *Social Work* 44(1): 5–8.

Witkin, S. 2001. "Complicating Causes." *Social Work* 46(3): 197–201.

Witkin, S. 2013. "Qualitative Social Work: Back to the Future." *Qualitative Social Work* 12(6): 722–731.

Witkin, S. 2014. "Compromise and Consequences: On the Challenges of Dialogue." *Qualitative Social Work* 13(2): 184–186.

Witkin, S. L. 2000. "Noticing." *Social Work* 45(2): 101–104.

Wolpert, L., and A. Richards. 1997. *Passionate Minds: The Inner World of Scientists.* Oxford: Oxford University Press.

Woolgar, S. 1981. "Interests and Explanation in the Social Study of Science." *Social Studies of Science* 11: 365–394.

Zeira, A., and A. Rosen. 2000. "Unraveling Tacit Knowledge: What Social Workers Do and Why They Do It." *Social Service Review* 74(1): 103–123.

Index